Guidance for Designing Communications Systems

For Water Quality Surveillance and Response Systems

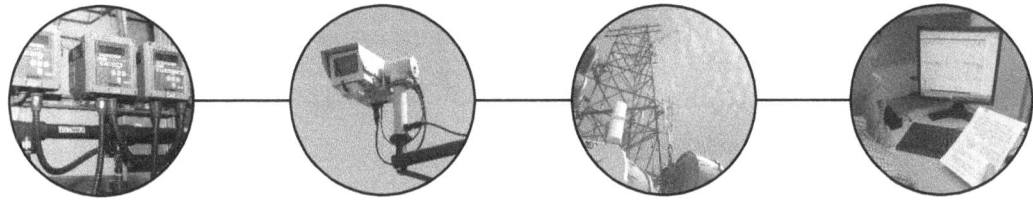

Disclaimer

The Water Security Division of the Office of Ground Water and Drinking Water has reviewed and approved this document for publication. This document does not impose legally binding requirements on any party. The information in this document is intended solely to recommend or suggest and does not imply any requirements. Neither the U.S. Government nor any of its employees, contractors or their employees make any warranty, expressed or implied, or assumes any legal liability or responsibility for any third party's use of any information, product or process discussed in this document, or represents that its use by such party would not infringe on privately owned rights. Mention of trade names or commercial products does not constitute endorsement or recommendation for use.

Questions concerning this document should be addressed to WQ_SRS@epa.gov or one of the following contacts:

Nelson Mix, PE, CHMM
EPA Water Security Division
1200 Pennsylvania Ave, NW
Mail Code 4608T
Washington, DC 20460
(202) 564-7951
Mix.Nelson@epa.gov

or

Steven C. Allgeier
EPA Water Security Division
26 West Martin Luther King Drive
Mail Code 140
Cincinnati, OH 45268
(513) 569-7131
Allgeier.Steve@epa.gov

Acknowledgements

The EPA Water Security Division would like to recognize the following individuals and organizations for their assistance, contributions, and review during the development of this document.

- Alan Lai, CH2M
- William H. Philips, CH2M
- Kenneth Thompson, CH2M
- Adam Haas, CSRA
- Goeffrey Brock, Philadelphia Water Department
- Ralph J. Rodgers, Philadelphia Water Department
- Michal Koenig, Qualcomm

Table of Contents

List of Figures

List of Tables

Abbreviations

2G	Second Generation
3G	Third Generation
4G	Fourth Generation
ADS	Anomaly detection system
AES	Advanced Encryption Standard
AMI	Automated Metering Infrastructure
dBm	Decibels referenced to one milliwatt
DNP3	Distributed Network Protocol 3
DSL	Digital Subscriber Line
EIRP	Equivalent Isotropically Radiated Power
EPA	United States Environmental Protection Agency
ESM	Enhanced Security Monitoring
FCC	Federal Communications Commission
GHz	Gigahertz
IEEE	Institute of Electrical and Electronics Engineers
IP	Internet Protocol
IT	Information Technology
kHz	Kilohertz
LAN	Local Area Network
LTE	Long-Term Evolution
MAC	Media Access Control
MAN	Metropolitan Area Network
MHz	Megahertz
MPLS	Multi-Protocol Label Switching
ORP	Oxidation Reduction Potential
OWQM	Online Water Quality Monitoring
P2P	Peer to Peer
POTS	Plain Old Telephone Service
PVC	Permanent Virtual Circuit
SCADA	Supervisory Control and Data Acquisition
SRS	Water Quality Surveillance and Response System
SVGA	Super Video Graphics Array
T1	T-Carrier 1 Line
TOC	Total Organic Carbon
TLS	Transparent LAN Service
VOIP	Voice-Over-Internet Protocol
VPN	Virtual Private Network
WAN	Wide Area Network
WiMAX	Worldwide Interoperability for Microwave Access
WPA2	Wi-Fi Protected Access 2

Section 1: Introduction

Data communications for a *Water Quality Surveillance and Response System (SRS)* is the means of transmitting data between remote monitoring sites and a drinking water utility's *control center*. Within an SRS, the *Enhanced Security Monitoring (ESM), Online Water Quality Monitoring (OWQM),* and *Automated Metering Infrastructure (AMI) components* typically include data generated by SRS equipment located at remote utility locations that is transmitted to the utility control center. When *alert* conditions are detected, utility personnel are notified via a *user interface* screen, email, or text message to initiate a response and, in some cases, utility personnel will transmit data to equipment at the remote sites as part of their investigation of SRS alerts (e.g., initiate collection of a water quality sample or position a video camera to observe an intruder). Drinking water utilities can use this document to evaluate alternatives, and design and implement an SRS communications system.

The overall process of evaluating alternatives consists of establishing evaluation criteria, identifying the available communications technologies for remote monitoring sites, and selecting a technology (or technologies). Evaluation and selection can be an iterative process depending on the available options, utility requirements, and budget. After a technology has been selected, evaluation criteria can be used to develop and implement the communications design.

A diagram of this process is provided in **Figure 1-1.**

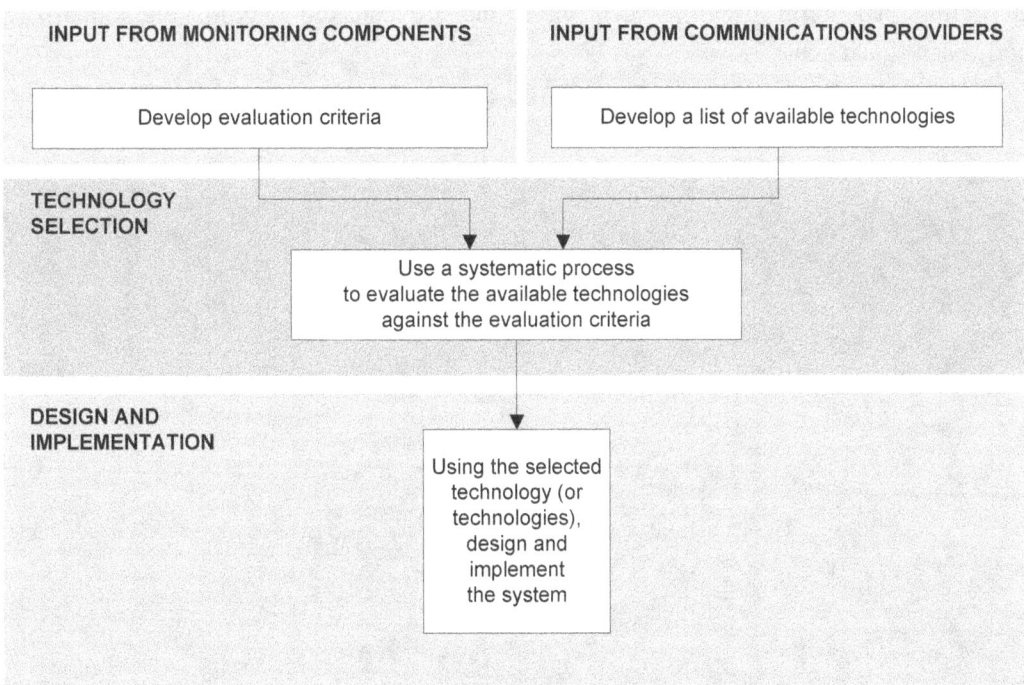

Figure 1-1. SRS Communications System Development Process

This document assumes a utility's *Information Technology (IT)* department would perform the technical tasks required in this document with input provided by utility personnel that are designing the remote monitoring equipment. A utility's *Supervisory Control and Data Acquisition (SCADA)* group may also be involved if the utility is considering leveraging or expanding the existing SCADA system for connectivity to remote monitoring equipment. The Information Technology and SCADA groups' involvement is essential to ensure that their policies and *constraints* are considered at each step. Furthermore, a utility's *wireless communications* group may be included in this process if wireless

1

technologies are being considered. Lastly, a utility's contracts administration group may be involved if the utility has a contract with a ***communications provider***.

If a utility does not have personnel with communications expertise, a communications consultant may be used to determine the utility's data communication requirements, evaluate and select potential communications providers, and develop a cost-effective solution. Furthermore, communications consultants can provide design services and implementation oversight if a utility-owned communications network is preferred.

This document is organized as follows:
1. **Section 2** describes the process of developing evaluation criteria.
2. **Section 3** describes commonly available communications technologies.
3. **Section 4** provides an overview of SRS component requirements for communications systems.
4. **Section 5** describes the process of selecting a communications system.
5. **Section 6** includes design and implementation guidance.
6. **Section 7** describes innovation in communications systems.
7. **Resources** presents a comprehensive list of documents, tools, and other resources cited in this document, including a summary and link to each resource.
8. **References** provides detailed information about source material cited in this document.
9. **Glossary** presents definitions of terms used this document, which are indicated by bold, italic font at first use in the body of the document.

Section 2: Evaluation Criteria

Evaluation criteria for a communications system are typically based on its data transmission characteristics, and operations and *maintenance* attributes. This section describes common overarching criteria including *extent of use, data transmission rate, security, reliability, distance, lifecycle cost,* and *coverage.* For each criterion, the description includes technical information and, where applicable, considerations for the criterion if the technology is utility-owned or provided by a third party (e.g., the local telephone company, cable television provider, or cellular carrier).

Criteria can be added, revised, or deleted based on utility needs and constraints. Examples of additional criteria include a utility's level of expertise with a technology and compatibility with existing communications protocols and *data management* equipment.

2.1 Extent of Use

Extent of use is a measure of acceptance by utilities that reflects the degree of confidence that the water sector has in the technology. This metric also measures the likelihood that vendors of the technology have experience working on water utility projects. A utility should research a technology's extent of use by surveying other utilities that use it or contacting a communications consultant that is knowledgeable about current communications methods.

Obsolete technologies should be avoided because *hardware* and provider availability could be limited and potentially costly. Be aware that some vendors try to sell older inventory. Typically, wireless systems are designed for 7 to 10 years of use before becoming obsolete, although legacy systems can be maintained using parts from inventory and third-party resellers.

However, emerging technologies often experience unanticipated issues that could require significant troubleshooting effort. The first year of service is usually covered under warranty, so issues during start-up and *commissioning* should be resolved at no cost to the end user.

2.2 Data Transmission Rate

A data transmission rate is the maximum, instantaneous rate of data that a technology can transmit. At a given site, the data transmission rate should be greater than the combined *data generation rate* of all equipment for proper communications system performance. This metric is typically expressed as data bits per second. Current technologies typically have data transmission rates that are in the gigabits, megabits, or kilobits per second range. Furthermore, the technology should be capable of accommodating the *data packet size* of all equipment and have a data *latency* that is suitable for the SRS application. Data generation rate, packet size, and latency are further described below.

2.2.1 Data Generation Rate

A data generation rate is the instantaneous rate of data produced by a piece of SRS equipment, such as a chlorine analyzer or video camera. The data generation rate of each type of equipment to be used must be estimated, and the quantities of each type of device should be determined. If the equipment type is already being used, data generation rates can be estimated by obtaining usage records from the utility department that manages the existing communications system or the service provider. For new equipment types, or if usage records are not available, a small-scale pilot using a representative sampling of new equipment under typical operating conditions can be used to empirically determine the equipment's data generation rates using network monitoring hardware and *software.* Utilities can also reach out to vendors or other utilities that have experience with the new equipment and evaluate the information from those programs.

3

If pilot testing is not preferred or other utility information is not available, data generation rates can be estimated using the calculations described in the Appendix.

The data generation rate for some equipment may be adjustable (i.e., a video camera's resolution may be reduced for a lower data generation rate), although operational needs should be kept in mind when reducing an equipment's data generation rate. Conversely, the impact on overall network performance should be considered when increasing an equipment's data generation rate.

2.2.2 Data Packet Size

SRS equipment will usually combine its generated data bits into a package of multiple bits called a data packet, which is transmitted to its destination. After the equipment receives an acknowledgement from the destination that the packet has been successfully received, the next data packet is sent. Data packet size is important when considering communications technologies that can experience poor signal conditions (e.g., wireless technologies) because a smaller data packet has a greater chance of successful transmission, although using smaller data packets requires more time for transmitting large amounts of data.

2.2.3 Latency

Latency is a metric related to data transmission rate and is defined as the amount of time between when a packet enters one end of a link and emerges from the other end. Latency is sometimes presented as "*round-trip latency*," which is the time between a packet entering one end of a link and an acknowledgment is received back at the same end of the link. Latency is a function of data transmission rate, hardware and software processing time, and the time required to establish the link.

To avoid timeout errors, software applications should be configured with conservative waiting periods when expecting data from communications links with long latencies. For most wired applications latency is not an issue, although relatively long latencies can be experienced with some wireless technologies. For most third-party provided technologies, a latency of a few seconds is readily achievable and should be adequate for SRS applications. However, latency will need to be carefully considered if near-instantaneous forms of communications such as Voice-Over-Internet Protocol (VOIP) will be included on the SRS communications link. Most third-party providers offer low latency options for an additional cost.

2.3 Security

Security measures are needed to ensure that the data transmitted between the remote sites and the control center cannot be viewed or altered by unauthorized personnel. Technologies that are not exposed to the Internet and provide data encryption integrated into the technology are generally regarded as being secure. For communications systems provided by a third party, it is important to determine if the methods of communication use the Internet or if there are options to use communications paths that are isolated from the Internet. Providers may charge an additional fee for an isolated connection. Security is discussed further in Section 6.4.

2.4 Reliability

Reliability means data reaches its destination completely, uncorrupted, and in the order it was sent. This criterion reflects the frequency and duration of periods of unacceptable performance for a technology. For a given communications technology, reliability can be reduced when the physical media for transmission is disrupted, for example, a cable break for wired applications or an obstruction in the signal path for wireless.

The reliability of a third-party service can also depend on the type of maintenance arrangement that is in place. For a *"best-effort" service*, providers are not obligated to maintain the subscription service level or pay penalties whenever the actual service level falls below the subscription service level. Additionally, the providers' responsiveness for repairing outages will vary, depending on the priority of other issues being addressed.

However, most *third-party provided services* can include a *service-level agreement* that defines *uptime* and reliability requirements, the response time for the provider to address unscheduled outages, and financial penalties when outages exceed the thresholds set in the agreement. An acceptable level of reliability for each SRS component should be balanced against cost when negotiating the terms of the service-level agreement with the provider.

2.5 Distance

Distance is the maximum extent that a technology can reliably transmit a packet of data without the aid of a signal repeater or amplifier. This measure should be carefully considered when evaluating wireless technologies because the distance between a remote site and a utility control center may exceed the maximum distance of the technology. This metric depends on the transmitter signal strength, receiver sensitivity, signal frequency, and physical media used for transmission (e.g., copper, fiber optic cable, or air).

2.6 Lifecycle Cost

The lifecycle cost for a communications system is an important criterion that allows the implementation and *operations and maintenance costs* for communications technologies to be compared. *Implementation costs* will vary greatly among technologies depending on the required hardware and infrastructure, and the effort needed to install, configure, and commission the system. Operational costs for third-party communications systems include recurring *provider fees,* which are further described below. However, for utility-owned systems, operational costs may be relatively small. Maintenance costs will depend on the type of agreement with the third-party provider or will rely on internal staffing needs for utility-owned systems. Additional discussion on maintenance costs is provided below. More details on lifecycle cost can be found in *Framework for Comparing Alternatives for Water Quality Surveillance and Response Systems*.

2.6.1 Provider Fees

For third-party provided communications systems, wired technologies are typically priced by data transmission rate, while data generated per month is usually the cost basis for cellular service. Data latency, service-level agreements, maintenance agreements, and security requirements can also factor into the provider's fees.

2.6.2 Maintenance

Maintenance is the cost and level of effort required to sustain the operation of a communications system. Maintenance can be performed by the third-party provider or by utility personnel.
For a third-party communications system, maintenance can be performed as part of best-effort or service-level agreements as described in Section 2.4. Maintenance can also be performed on a fully managed or unmanaged basis. An unmanaged arrangement requires that the utility maintain the on-site equipment and wiring as well as

> **HELPFUL HINT**
>
> Most communications vendors offer annual maintenance contracts, although some will provide maintenance services on a time-and-materials basis. The anticipated amount of repairs needed over a given year will determine which arrangement is more cost effective.

diagnose outages that are not caused by utility-owned equipment. Such an arrangement may be suitable for utilities that have personnel with expertise in maintaining communications and networking equipment.

For a fully managed solution, the provider owns, configures, monitors, and maintains the equipment needed for the service to function. This is an appealing option for utilities that do not have personnel with wide-area-network (WAN) management expertise or do not want to purchase WAN maintenance equipment. However, the utility may have a higher recurring cost if a fully managed solution is chosen.

2.7   Coverage

Coverage is the geospatial availability of technologies and must be considered when determining the communications options for each site. Coverage will vary by location and technology. Multiple communications technologies may be needed to provide coverage to all SRS sites, especially for large utilities that include SRS sites extending beyond the reach of any single provider or are unfeasible to include in a utility-owned wireless system. In general, the coverage of emerging technologies will expand as providers upgrade their infrastructure, while the coverage of older technologies will shrink as providers decommission aging infrastructure and phase out obsolete systems.

Section 3: Communications Technologies

This section provides a summary of commonly available technologies including wired technologies, which require some form of solid physical media for transmitting data (e.g., copper wiring or fiber optic cabling), and wireless technologies, which transmit data via electromagnetic signals that travel through air. The wired technologies described are the Plain Old Telephone Service (POTS), Digital Subscriber Line (DSL), T-Carrier 1 Line (T1), Frame Relay, Multi-Protocol Label Switching (MPLS), Transparent Local Area Network (LAN) Service (TLS), and utility-owned fiber optic. The wireless technologies described are digital cellular and utility-owned wireless communications.

Table 3-1 includes the most common technologies offered and qualitative ratings against the previously described evaluation criteria (except for coverage, which is site specific). The qualitative ratings listed in the table are based on input from subject matter experts and should be refined for a utility's specific application based on information from communications providers. Additional information on wired and wireless technologies is provided after Table 3-1 and includes detailed descriptions of the wired and wireless technologies and an explanation of each rating.

3.1 Wired Technologies

Most wired technologies are capable of relatively high-speed data transmission suitable for SRS data needs, with the exception of leased POTS lines that operate at speeds of 56 kilobits per second. Wired technologies typically cost more than wireless technologies due to the need for underground or overhead cabling. If underground conduit is required, incorporating the trenching into a capital project with other excavation can reduce the overall installation costs. Furthermore, wired technologies have a high level of reliability as long as the cabling is not physically damaged. Wired technologies are typically provided by a third party such as a telephone or cable television company, although in some cases a utility may own fiber optic connections between its facilities.

Wired technologies such as T1, TLS, MPLS, and utility-owned fiber optic are often used for data communications between major utility facilities, although wireless technologies are typically used for communicating with facilities where *wired communication* is not available or practical (e.g., AMI or SRS equipment located at remote utility facilities).

The following subsections present a description and detailed performance rating for each of the wired communications technologies presented in Table 3-1.

Table 3-1. Commonly Available Communications Technologies

	Communication Technology	Extent of Use	Data Tr. Rate	Security	Reliability	Distance	Installation Cost	Provider Fees	Maintenance
Wired	**Plain Old Telephone System (POTS):** POTS is the basic form of wired voice communication. A conventional modem can be used over POTS for data communication, but is limited to 56 kilobits per second without data compression.	○	○	●	◐	●	●	◐	◐
	Digital Subscriber Line (DSL): DSL uses existing POTS infrastructure for data transmission between facilities via the Internet, although some providers offer a private network option at additional cost. DSL is capable of transmission rates of up to 5,000 kilobits per second to the end user and up to 768 kilobits per second from the end user.	◐	●	○	◐	●	●	◐	◐
	T-Carrier 1 (T1) Line: A T1 line is a dedicated point-to-point data connection between facilities that is capable of transmission rates up to 1.54 megabits per second.	◐	●	●	●	●	○	○	●
	Frame Relay: To the end user, frame relay appears to be a dedicated point-to-point data connection up to 1.5 megabits per second, similar to a T1 line. However, providers vary the size and routing of frame relay data packets to optimize usage of their infrastructure, resulting in a reduction in costs relative to that of T1 lines.	◐	●	●	●	●	○	◐	●
	Multi-Protocol Label Switching (MPLS): This newer technology is replacing T1 and frame relay connections and capable of transmission rates up to 622 megabits per second	●	●	●	●	●	◐	◐	●
	Transparent LAN Service (TLS): Also called "Metro Ethernet," TLS is an emerging technology that provides an Ethernet data transmission rate connections between facilities of 10, 100, or 1000 megabits per second.	◐	●	●	●	●	●	◐	●
	Utility-Owned Fiber Optic: This dedicated point-to-point data connection between facilities is capable of transmission rates up to 10 gigabits per second.	◐	●	●	●	●	○	●	○
Wireless	**Digital Cellular:** Digital cellular uses wireless transceivers to connect to a provider's cellular network for data transmission. The cellular technologies, third generation (3G) and fourth generation (4G), have transmission rates of up to 800 kilobits per second and 10 megabits per second, respectively. Upload and download data transmission rates are often asymmetric with upload rates being lower.	◐	●	◐	◐	●	●	◐	●
	Utility-Owned Wireless: Utility-owned wireless uses utility equipment and infrastructure for data transmission over unlicensed or licensed frequency bands. Transmission rates vary widely depending on the modulation technology and frequency band (9.6 kilobits per second for low-speed, narrowband technologies and up to 7 gigabits per second for high-speed Wi-Fi). This category also includes microwave technologies.	●	●	◐	◐	◐	○	●	○

Attribute Key: ● Strong ◐ Moderate ○ Weak

3.1.1 Plain Old Telephone Service

Also called Publically Switched Telephone Network, POTS technology consists of analog telephone lines and is a basic offering of the local telephone company. The quantity of POTS connections are decreasing in favor of faster digital technologies for data transmission and cellular telephones for voice communication. Thus, many providers are reviewing their level of support for POTS technology. However, POTS is one of the few technologies that will work during a power outage and often plays a critical role in emergency management operations for voice communications. A description of each performance rating for this technology is provided in **Table 3-2**.

Table 3-2. Plain Old Telephone System (POTS) Performance Ratings

Performance Criteria	Rating	Description/Notes
Extent of Use	O	For data transmission, POTS is being replaced by digital technologies such as DSL, Frame Relay, MPLS, and Metro Ethernet.
Data Transmission Rate	O	Limited to 56 kilobits per second, which is slow when compared to digital technologies, but may be adequate for ESM applications with low data generation rates, such as those without video.
Security	●	Not exposed to the Internet.
Reliability	◐	"Best effort" service is typical for POTS. Connections are designed for voice communications and signal noise, which may be acceptable for voice communications, can reduce the quality of data transmissions.
Installation Cost	●	In terms of equipment costs, POTS modems are inexpensive when compared to that required by other wired technologies. Installation costs are also relatively low since existing telephone lines can be used, which are available at most facilities.
Provider Fees	◐	Unless a utility has their own POTS system, monthly fees to the telephone provider are required for POTS, but they are generally less than fees for digital technologies.
Maintenance	◐	Maintenance is generally handled by the provider on a "best effort" and unmanaged basis.
Attribute Key: ● Strong ◐ Moderate O Weak		

3.1.2 Digital Subscriber Line

DSL technology uses existing POTS infrastructure for digital data transmission. Like POTS, DSL can be a viable option for remote utility facilities that may not have any existing wired infrastructure other than an analog telephone line. DSL connections generally provide Internet service to the customer and provide data communications between facilities. DSL is usually provided by the local telephone company. A description of each performance rating for this technology is provided in **Table 3-3**.

Table 3-3. Digital Subscriber Line Performance Ratings

Performance Criteria	Rating	Description/Notes
Extent of Use	◑	DSL is an option where POTS exists.
Data Transmission Rate	●	Capable of up to 5,000 kilobits per second from the remote site and up to 768 kilobits per second to the remote site. The data transmission rate depends on the DSL technology, line conditions, and service-level implementation.
Security	○	Most DSL connections use the Internet for connectivity. However, some providers offer a solution that is isolated from the Internet, but at significant additional cost.
Reliability	◑	"Best-effort" service is typical for DSL.
Installation Cost	●	In terms of equipment costs, DSL modems are more expensive than those used with POTS, but generally less than that of other wired technologies. Installation costs are relatively low because existing telephone lines can be used for DSL, which are available at most facilities.
Provider Fees	◑	Has recurring costs, but not as much as those required by the T1, Frame Relay, TLS, and MPLS technologies.
Maintenance	◑	Maintenance is generally handled by the provider on a "best effort" and unmanaged basis.
Attribute Key: ● Strong ◑ Moderate ○ Weak		

3.1.3 T-Carrier 1 Line

T1 circuits are typically offered by the telephone company and provide point-to-point connections between facilities. T1 lines can be used to connect multiple facilities to form a WAN. A description of each performance rating for this technology is provided in **Table 3-4**.

Table 3-4. T1 Line Performance Ratings

Performance Criteria	Rating	Description/Notes
Extent of Use	◒	Although used widely by many utilities over the past decade, most T1 lines have been or are being converted to MPLS or Metro Ethernet (TLS) service.
Data Transmission Rate	●	Data transmission rates of up to 1.54 megabits per second. A full T1 connection is made up of twenty-four, 64 kilobits-per-second channels plus 8 kilobits per second of communications overhead. Fractional T1 connections are also available if transmission rates of less than 1.54 megabits per second are required.
Security	●	T1 lines are usually dedicated connections between facilities and are not exposed to the Internet.
Reliability	●	Service-level agreements are typical for T1 connections.
Installation Cost	○	In terms of equipment costs, T1 hardware is relatively expensive when compared to that of POTS and DSL. Installation is also relatively expensive because T1 lines typically require upgrades to the building wiring.
Provider Fees	○	Monthly T1 fees can be relatively high when compared to POTS and DSL. Costs can be reduced by purchasing fractional T1 connections if a data transmission rate less than 1.54 megabits per second is acceptable.
Maintenance	●	Maintenance is generally handled by the provider based on a service-level agreement and a fully managed basis.
Attribute Key: ● Strong ◒ Moderate ○ Weak		

3.1.4 Frame Relay

This technology is a less-expensive alternative to T1 lines for interconnecting facilities to form a WAN and is typically provided by the telephone company. Unlike T1, which uses dedicated point-to-point connections between facilities, frame-relay circuits connect each facility to the nearest node in the provider's network. The provider's frame-relay service then routes inter-facility data through the provider's network infrastructure to the proper destination. To the end user, a frame-relay connection, sometimes called a permanent virtual circuit (PVC), appears to be continuous and dedicated, but the provider can vary the size and routing of data packets to optimize usage of its infrastructure, reducing costs. A description of each performance rating for this technology is provided in **Table 3-5**.

Table 3-5. Frame Relay Performance Ratings

Performance Criteria	Rating	Description/Notes
Extent of Use	◒	Although used widely by many utilities over the past decade, most frame relay connections have been or are being converted to MPLS or Metro Ethernet (TLS) service. In some areas, frame relay may no longer be available.
Data Transmission Rate	●	Similar to that of a T1 line, up to 1.5 megabits per second.
Security	●	Usually not exposed to the Internet.
Reliability	●	Service-level agreements are typical for frame-relay connections.
Installation Cost	○	Similar to that of a T1 line, frame-relay equipment is relatively expensive when compared to POTS and DSL hardware. Installation is also relatively expensive because frame-relay lines typically require upgrades to the building wiring.

Performance Criteria	Rating	Description/Notes
Provider Fees	◒	Unlike T1, which provides a dedicated connection between endpoints, frame relay is packet-switched, allowing providers to more efficiently use their network capacity, and generally less expensive than dedicated T1 lines. Recurring costs for frame relay users are typically based on the committed information rate, which is a data transmission rate that the provider guarantees will be available, and the amount of data transmitted over a specified period of time.
Maintenance	●	Maintenance is generally handled by the provider based on a service-level agreement and a fully managed basis.
Attribute Key: ● Strong ◒ Moderate ○ Weak		

3.1.5 Multi-Protocol Label Switching

MPLS service, available from third-party providers such as the local telephone company, is a high-capacity metropolitan area network (MAN) protocol used to route traffic across the service provider's network to connect multiple nodes. MPLS encapsulates traffic of various protocols and uses labels to route the encapsulated traffic between nodes. Nodes can be connected to the MPLS using a range of common protocols including T1, frame relay, and DSL. A benefit to MPLS technology is that it can be used to transmit data from a user's DSL transceivers over a connection that is not exposed to the Internet, which is more secure, whereas most traditional DSL providers use the Internet to provide connectivity as previously described in Section 3.1.2. A description of each performance rating for this technology is provided in **Table 3-6**.

Table 3-6. Multi-Protocol Label Switching Performance Ratings

Performance Criteria	Rating	Description/Notes
Extent of Use	●	Becoming more common and replacing older technologies such as T1 and frame relay.
Data Transmission Rate	●	Data transmission rates of up to 622 megabits per second.
Security	●	Most MPLS connections use the provider's internal network and are not exposed to the Internet.
Reliability	●	Service-level agreements are typical for MPLS connections.
Installation Cost	◒	In terms of equipment costs, MPLS equipment is relatively expensive when compared to POTS and DSL hardware, but is less than T1 or Frame Relay equipment. Installation is also relatively expensive because MPLS lines typically require upgrades to the building wiring.
Provider Fees	◒	Varies with bandwidth. Generally more expensive than TLS services, but less than T1 and Frame Relay services.
Maintenance	●	Maintenance is generally handled by the provider based on a service-level agreement and a fully managed basis.
Attribute Key: ● Strong ◒ Moderate ○ Weak		

3.1.6 Transparent LAN Service

TLS, also called Metro-Ethernet service, is available from third-party providers, typically the local telephone or cable television company, for interconnecting facility networks such that they appear to the user as one contiguous network. A description of each performance rating for this technology is provided in **Table 3-7**.

Table 3-7. Transparent LAN Service Performance Ratings

Performance Criteria	Rating	Description/Notes
Extent of Use	◓	Newer than MPLS, TLS has been replacing T1 and frame relay connections. TLS is available in many large metropolitan areas, but may not be offered in suburban and rural areas.
Data Transmission Rate	●	Data transmission rates of 10, 100, or 1000 megabits per second.
Security	●	TLS connections usually use the provider's internal network and are not exposed to the Internet.
Reliability	●	Service-level agreements are typical for TLS connections.
Installation Cost	●	In terms of equipment costs, TLS equipment is relatively expensive when compared to POTS and DSL hardware, but is less than MPLS. Equipment costs are similar to that of T1. Installation is also relatively expensive because TLS lines typically require upgrades to the building wiring.
Provider Fees	◓	Varies with bandwidth. Generally less expensive than MPLS services.
Maintenance	●	Maintenance is generally handled by the provider based on a service-level agreement and a fully managed basis.
Attribute Key: ● **Strong** ◓ **Moderate** ○ **Weak**		

3.1.7 Utility-Owned Fiber Optic

If a utility owns fiber optic cables between facilities, it is likely that this cable can be used to transmit data from SRS equipment, including equipment with high data generation rates. Fiber optic cabling between facilities may also be leased from a local service provider such as a cable television company. Because most fiber optic cables have multiple pairs of conductors, a spare pair may be available. Alternately, existing network connections over fiber optic could be used to transmit large amounts of data quickly depending on the data generation rate of existing *datastreams.* When installing pipelines between facilities, a utility might consider installing a conduit with fiber optic cabling in the same trench. The benefits of fiber optic cabling, which include high data transmission rates and long distances between repeaters, can offset the relatively small cost of installation in a trench already dug for a pipeline, especially since the majority of construction cost is from excavation. A description of each performance rating for this technology is provided in **Table 3-8**.

Table 3-8. Utility-Owned Fiber Optic Performance Ratings

Performance Criteria	Rating	Description/Notes
Extent of Use	◔	This is used by some utilities, but not as widespread as T1.
Data Transmission Rate	●	Capable of data transmission rates up to 10 gigabits per second
Security	●	Fiber lines are dedicated point-to-point connections, so very secure.
Reliability	●	Properly installed fiber optic lines are not susceptible to electromagnetic interference and usually have a high level of uptime. The other wired technologies are copper-based and use shielding and/or conductor twisting to minimize electromagnetic interference. However, fiber optic cabling can be susceptible to inadvertent breakage by unsuspecting excavators.
Distance	●	Fiber optic connections can transmit data up to 20 miles between repeaters.
Installation Cost	○	In terms of equipment costs, fiber optic equipment is relatively expensive when compared to POTS and DSL hardware, but usually less than T1 or frame relay equipment. Installing fiber optic lines between facilities can be costly, requiring trenching or overhead infrastructure.
Provider Fees	●	If a utility owns its fiber optic connection, there are no monthly fees. If leased from a third party, then provider fees will apply.
Maintenance	○	The utility is responsible for all maintenance.
Attribute Key: ● Strong ◔ Moderate ○ Weak		

3.2   Wireless Technologies

A key advantage of wireless technologies is that hardwired infrastructure is not required, saving the cost of trenching, poles, and cabling. However, an unobstructed path between the sending and receiving antenna is strongly recommended for the successful transmission of a wireless signal. Lower frequency signals can tolerate some obstructions, depending on the application and physical characteristics of the blockage. Foliage is a typical obstruction because the water content in leaves absorbs the energy from a wireless signal. Heavy fog can also reduce the strength of a wireless signal.

Wireless technologies can be owned and maintained by the utility or provided by a cellular carrier. For a utility-owned wireless system, the utility designs, implements, operates, and maintains the wireless network. For a cellular system, the utility provides input at the design and implementation stages, but the operation and maintenance of the network is the primary responsibility of the cellular provider, and the utility will incur monthly usage fees. Such trade-offs can be evaluated using lifecycle costs as previously described in Section 2.6.

Wireless technologies have been rapidly evolving. Newer adaptive modulation schemes allow wireless technologies to transmit more data per hertz of frequency bandwidth and be more resilient against interference from other wireless

> **CASE STUDY**
>
> A large utility experienced numerous communications outages that lasted multiple weeks after fiber optic cables were damaged in areas with heavy construction. Over the evaluation period, this utility found that cellular communications experienced considerably less downtime.

transmitters when compared with that of older technologies. Examples of adaptive modulation technologies include Orthogonal Frequency Division Multiplexing and Quadrature Amplitude Modulation, which can both be used by utility-owned wireless systems. These adaptive modulation schemes have also been used in digital cellular systems as the core technologies behind *Worldwide Interoperability for Microwave Access (WiMax)* and *Long-Term Evolution (LTE)*, which are cellular methods of transmitting data at 4G rates.

The following subsections present a description and detailed performance rating for each of the wireless communications technologies presented in Table 3-1.

3.2.1 Digital Cellular

Applications of cellular-based, *machine-to-machine communications* have grown exponentially as part of the *Internet of Things* revolution, which anticipates over 50-billion connected devices worldwide by 2020 (Menon, 2015). The momentum of this technology makes it an appealing option for SRS applications because of broad coverage, relatively low installation costs, falling monthly costs, increased data transmission rates, and improved reliability due to the rigorous testing required prior to being authorized for use. However, utilities should avoid second generation (2G) cellular (also known as 1x) because this technology is being phased out and will not be supported in the near future.

CASE STUDY

A large utility deployed a 2G cellular system for its ESM and OWQM components, and found its performance for transmitting ESM incident-driven video to be marginal, at best. However, the 2G service performed reliably for the OWQM component.

The 2G cellular was eventually replaced with T1 and DSL connections at ESM sites to improve video performance and later at OWQM sites when the cellular carrier discontinued 2G service.

One or more cellular carriers have coverage in all metropolitan and suburban areas with an increasing number of remote and rural areas also receiving cellular service. However, there will be localized areas that may not receive cellular coverage.

The widespread coverage of cellular service means that antenna masts and repeater sites are not required to obtain an adequate signal at remote SRS sites, which reduces installation costs. However, the user typically purchases the transceivers, antennas, and cabling. If a facility has existing cellular connectivity, it may be possible to avoid installation costs and enable additional bandwidth by opting for a higher-rate data plan, assuming that the existing cellular hardware is capable of transmitting data at the higher rate. If not, hardware upgrades may be required.

A description of each performance rating for this technology is provided in **Table 3-9**.

Table 3-9. Digital Cellular Performance Ratings

Performance Criteria	Rating	Description/Notes
Extent of Use	◕	Becoming more prevalent as wireless providers expand their networks and increase network capacity.
Data Transmission Rate	●	3G and 4G are up to 800 kilobits per second and 10 megabits per second, respectively, which is suitable for applications with high data generation rates, such as video traffic. Areas with only 3G coverage typically have data transmission rates between 500 and 800 kilobits per second depending on the carrier. Each carrier's implementation of 3G and 4G technologies differs slightly, so on-site testing of a carrier's actual data transmission rate may be needed to determine suitability for a utility's application. 5G technologies are anticipated to be available in 2020 with transmission rates exceeding 1 gigabit per second.
Security	◕	The link between the cellular provider's network and the utility's network may use the Internet. If this is the case, a Virtual Private Network (VPN) connection is strongly recommended to maintain data security and integrity between the provider's and utility's networks. Alternately, some cellular providers offer a direct connection between their network and the utility's network that is isolated from the Internet for an additional fee. VPN connections from the cellular transceiver to the provider's cell tower are also recommended to protect against "man-in-the-middle" attacks to the wireless link. 4G cellular chipsets have built-in encryption, further securing data transmitted over this technology. However, this technology may be subject to malicious interference.
Reliability	◕	Subject to congestion and signal level-related interruptions, although the frequency of such outages continues to decrease as cellular carriers improve the speed and coverage of their infrastructure. Usually a "best effort" service with no service-level agreement. Cellular companies continue to build towers in urban and suburban areas to reduce the number of dropped connections and in rural and remote areas for broader coverage. Also, cellular companies typically design their towers to withstand weather emergencies and power outages.
Distance	●	Approximately 2 miles between the transceiver and nearest cell tower, depending on antenna location and obstructions. Directional antennas and, in some cases, amplifiers can be used to increase a cellular transceiver's effective range.
Installation Cost	●	The cellular provider usually recommends the transceivers and antennas that the utility must purchase to connect to the provider's network, so equipment costs will vary. When cell towers are nearby, no antenna towers are needed and installation is relatively inexpensive. In most cases, the antenna can be located near street level and will not require a mast as cellular networks are designed for connectivity at ground level.
Provider Fees	◕	Monthly fees are required, but they are not as expensive as wired technologies such as T1, frame relay, or TLS lines. Cellular providers charge recurring fees for use of their networks. Data plans for a utility would be tiered based on an allotment of data transferred per month, similar to consumer data plans. Check with local and nationwide carriers in the utility's service area for their data plans and rates.
Maintenance	●	Maintenance is generally handled by the provider on a "best effort" and unmanaged basis.
Attribute Key: ● Strong ◕ Moderate ○ Weak		

3.2.2 Utility-Owned Wireless

Utility-owned wireless systems have been used by utilities for communicating with remote sites for decades. Design of a wireless network or of a new node on an existing network requires a wireless signal path analysis to determine antenna heights and whether repeaters will be needed, especially in areas with a contoured topography and when the distance between transceivers is approaching the practical range of the technology. Depending on the signal path, it may be possible to locate a repeater on a utility-owned structure. However, in many cases, the repeater may need to be located on a privately-owned structure, which could add complexity to the installation. Software is available for desktop-level preliminary path analysis, but an on-site analysis of the wireless sites and the path in between is strongly recommended to locate potential obstructions that are not adequately represented in the terrain data used by the software.

The effective distance of a wireless transmitter is a function of the frequency, with lower frequencies capable of transmitting longer distances. The frequency-distance relationship applies to all wireless technologies, including digital cellular. However, in practice this relationship is only a consideration for utility-owned wireless because digital cellular frequencies are determined by the provider. Utilities can use parabolic and other directional antennas to focus wireless signals and offset the attenuation effects of higher frequencies. Furthermore, utilities should consult the manufacturer's literature for acceptable distances between wireless transceivers, although many wireless installation professionals suggest dividing the manufacturer's stated maximum distances between wireless transceivers by a factor of two to be conservative.

> **HELPFUL HINT**
>
> Permitting for towers and antennas has been difficult in many communities due to aesthetic issues and, in some cases, public health concerns regarding electromagnetic wave exposure. While the reasonableness of the public health concern is still questionable, it is often raised by community groups opposing a project.

Utility-owned wireless networks can use unlicensed or licensed frequency bands. Unlicensed frequency bands are relatively inexpensive to implement, but they have limits on transmission power and may be subject to occasional interference. Utilities with applications that need higher levels of transmission power and minimal interference can opt to use licensed bands, which can require more effort and cost to implement. A description of both approaches follows.

Unlicensed Frequency Bands

Although utility-owned wireless networks can be implemented relatively quickly in unlicensed frequency bands, interference can be an issue depending on the quantity, frequency, signal strength, and direction of nearby transmitters. Ideally, sources of interference should be identified during the design phase by conducting an on-site spectrum analysis at each proposed site so that a mitigation approach can be developed before installation. If a potential source of interference is found, the utility may need to consider installing directional antennas or changing the configuration of the transceiver to ignore the interfering frequencies. Although a spectrum analysis may not reveal any sources of interference during the design phase, it is possible that a transceiver may be constructed in the future that could impact the utility-owned wireless network. Web-based tools are available for conducting a preliminary search for wireless transmitters near a proposed location.

> **HELPFUL HINT**
>
> Consult Code of Federal Regulations 47 part 15 for the Federal Communications Commission (FCC) rules that govern unlicensed wireless operation. Power limitations are discussed in Subpart C, and managing interference is described in Subpart A. Although users can utilize an entire unlicensed frequency band without coordinating with other users of the band, Subpart A requires that a wireless transmitter operating in an unlicensed frequency band that is causing *harmful interference* must cease operation until the condition causing the harmful interference has been corrected. Harmful interference is defined as transmissions that endanger the functioning of a radio navigation service or of other safety services or seriously degrades, obstructs or repeatedly interrupts a wireless communications service operating in accordance with FCC rules.

The data transmission rate of unlicensed transceivers depends on the modulation scheme used. Unlicensed transceivers that use adaptive modulation schemes such as Quadrature Amplitude Modulation and Orthogonal Frequency Division Multiplexing are currently available. These transceivers provide data transmission rates above 100 megabits per second for point-to-point applications, which are dedicated communications links between two transceivers. For point-to-multipoint applications, which consist of a base transceiver (access point) that communicates with multiple remote transceivers (subscribers), adaptive modulation schemes provide data transmission rates of up to 54 megabits per second. Therefore, these types of transceivers should be considered for video and other bandwidth-intensive applications. Older modulation schemes such as direct sequence spread spectrum and frequency hopping spread spectrum are still available, but typically do not provide data transmission rates above 1 megabit per second, although some manufacturers claim rates of up to1.5 megabits per second. Such transceivers are better suited for ESM applications without video and OWQM stations monitoring simple datastreams.

Most transceivers offer Advanced Encryption Standard (AES) 128 encryption and other security features. However, utility wireless communication security requirements often require the use of additional security hardware (e.g., firewall, content filtering device, cyber-intrusion detection device) to provide acceptable data integrity and, when necessary, confidentiality.

Utilities typically have installed wireless networks using the following unlicensed frequency bands.

- **902-928 Megahertz (MHz):** This frequency band is the most widely used largely due to the longer range of transceivers that use this band. However, its widespread usage also makes interference more likely. Cellular companies often use a frequency just below 902 or just above 928 MHz and can be a significant source of interference due to their relatively high signal strength. In this frequency band, *Equivalent Isotropically Radiated Power (EIRP)*—a commonly used measure of emitted radio frequency power referenced to one milliwatt—is limited to 36 decibels referenced to one milliwatt (dBm).

- **2400-2500 MHz:** This frequency band is not as commonly used for utility communications as the 902-928 MHz band due to the shorter range of transceivers that use the 2400-2500 MHz band. A benefit to having fewer users is a lower potential for interference. However, the 2400-2500 MHz frequency range is one of the bands used by *Wi-Fi,* which can cause interference in areas with Wi-Fi usage. As in the 902-938 MHz band, EIRP for 2400-2500 MHz is limited to 36 dBm.

- **5725-5875 MHz:** This is the least used of the unlicensed bands. However, point-to-point transceivers and point-to-multipoint subscribers using directional antennas are allowed to transmit at 46 dBm (EIRP) in most cases. Users of this band sometimes employ multi-sector access points for increased data throughput for point-to-multipoint applications. A multi-sector access point consists of three or four directional antennas that cover 90 or 120 degrees each and a dedicated base transceiver connected to each antenna. The base transceivers are synchronized to form an access point that is capable of handling three to four times the amount of traffic that an access point equipped with a single base transceiver and omnidirectional antenna could have accommodated. However, similar to 2400-2500 MHz, the 5725-5875 MHz frequency range is one of the bands used by Wi-Fi, which can cause interference in areas with Wi-Fi usage.

Wi-Fi is an unlicensed wireless technology that conforms to the Institute of Electrical and Electronics Engineers (IEEE) 802.11 standard. It is currently available in five different versions, 802.11b, 802.11g, 802.11n, 802.11ac, and 802.11ad, which have data transmission rates of up to 11, 54, 450, 1350, and 7,000 megabits per second, respectively. Both versions 802.11n and 802.11ac can have up to three streams, and thus their capacities are 450 and 1,350 megabits per second, respectively. However, transmit powers are restricted well below those for the non-standards-based unlicensed alternatives discussed above, which significantly reduces outdoor Wi-Fi network ranges. Using standard omni-directional antennas, the "b," "g," and "ad" versions have a maximum outdoor range of approximately 300-400 feet, while the "n" and "ac" versions can extend outdoors beyond 800 feet. Longer ranges of ¼- to ½-mile outdoors can be achieved using "n" and "ac" equipment designed for outdoor applications and high gain panel antennas. Widespread consumer usage has driven down the cost of this technology; however the prevalence of Wi-Fi also increases the vulnerability of such a communications link to cyberattack. Industry-wide best practices such as Wi-Fi Protected Access 2 (WPA2) encryption, disabling broadcasts, and enabling Media Access Control (MAC) address filtering are strongly recommended to keep Wi-Fi connections secure.

Licensed Frequency Bands

The benefits of using a licensed wireless network is that mechanisms are provided for mitigating interference and the permissible amount of transmission power is generally higher than what is allowed for unlicensed transceivers. The data transmission rates and licensing process of licensed transceivers and mechanisms for managing interference vary from band to band. The details associated with some of the more common licensed frequency bands are described below.

- **3650-3670 MHz:** This licensed frequency band is for industrial use and can be used by utilities. All licenses for this band are free and nationwide, although a registration fee of approximately $300 is required. Licensees are required to examine the FCC's Universal Licensing System registration database before registering a station, and they must coordinate with other 3.65 gigahertz (GHz) band operators to avoid interference, as priority is not given to licensees who were first to deploy in any given service area. Licensees of stations suffering or causing harmful interference are expected to cooperate and resolve the problem by mutually satisfactory arrangements. However, the effectiveness of this cooperative approach has not yet been proven and, if unsuccessful, the resulting congestion may make this band less viable for critical communications. Adaptive modulation technologies are used in this band to provide higher data transmission rates, and, similar to the 5725-5875 MHz unlicensed band, users of the 3650-3670 MHz licensed band can use multi-sector access points and adaptive modulation technologies such as Quadrature Amplitude Modulation and Orthogonal Frequency Division Multiplexing to increase data throughput. However, the limitations on channel bandwidths together with the limited number of channels reduce data transmission rates and the total number of subscribers that can be effectively served in each sector.

  Transceivers complying with the WiMAX standard, IEEE 802.16e, are available in this band and can be used to build private data networks with multi-sector access points. Using these transceivers allows owners to tightly manage subscriber bandwidth and quality of service. These powerful tools could make it possible to build private metropolitan networks similar to digital cellular networks although, because of mobile transceiver transmit power limits and other restrictions, the private networks do not support mobile users very well.

- **4940-4990 MHz:** This licensed frequency band is for public safety applications. A single regional license for all channels in the band is granted to a public safety entity for a geographical area encompassing a legal jurisdiction. A regional plan serves to facilitate shared use in each region. Similar to the 3650-3670 MHz licensed band, users of the 4940-4990 MHz band can use multi-sector access points and adaptive modulation technologies to increase data throughput. However, the limitations on channel bandwidths, together with the limited number of channels, reduce data transmission rates and the total number of subscribers that can be effectively served in each sector.

- **6 GHz through 23 GHz:** These licensed microwave frequency bands are typically used for high-speed, backhaul communications by wireless Internet service providers in rural areas, cellular companies, and telecommunications companies. Transceivers that use microwave frequency bands transmit data at approximately 150-800 megabits per second and can also be used by utilities for SCADA backhaul, video surveillance, and other point-to-point applications.

- **Narrowband Bands:** Each of these bands encompasses much less spectrum than the licensed broadband bands discussed above, with some bands being only 3 MHz wide. Channel bandwidths are typically 6.25 or 12.5 kilohertz (kHz) as compared to broadband channel bandwidths, which are in multiples of 1 MHz. However, channels can be aggregated into 25 and, in some cases, 50 kHz channels with the exception of the Very High Frequency and Ultra High Frequency bands (150-174 MHz and 421-512 MHz), which are currently limited to 12.5 kHz and eventually will be reduced to 6.25 kHz. These bands fall into three basic categories: Primary Use, Secondary Use, and Auctioned. Performance characteristics, components, and rules associated with licensing and use of each band within each category can vary significantly.

o   In Primary Use bands, which include 928-929 MHz, 932-932.5 MHz, 941-941.5 MHz, and 952-958 MHz, fixed operations such as telemetry and SCADA are designated as primary users and are protected by regulations and licensing procedures from co-channel interference.

o   In Secondary Use bands, which include 150-170 MHz, 217-220 MHz and 450-470 MHz, fixed operations such as telemetry and SCADA are designated as secondary users and are required to resolve any interference issues with primary users, which are usually land mobile-transceiver users. Interference is a significant liability for fixed users in the Secondary Use bands because the primary users can move around anywhere inside the service area.

o   The third category, Auctioned, which includes 220-222 MHz, 901.B-901.7 MHz, 930-931 MHz, and 940-941 MHz, are bands that have been auctioned to a single licensee that then sells the channels in that band to users for use anywhere within a specific county. The auctioned bands are significantly more expensive than other licensed bands, offer more operational flexibility, and are less likely to experience interference problems as there is only one user in each county.

Knowledge of and experience with each band's characteristics and components, as well as careful coordination to avoid interference issues and adherence with FCC rules, are required to select and effectively use any of these bands. A description of each performance rating for this technology is provided in **Table 3-10**.

Table 3-10. Utility-Owned Wireless Performance Ratings

Performance Criteria	Rating	Description/Notes
Extent of Use	●	Unlicensed: Used widely by many utilities. Wi-Fi: Used widely by many utilities for office settings, but not as widely used for communicating to remote facilities. Licensed: Least used of wireless technologies.
Data Transmission Rate	●	Unlicensed: Using adaptive modulation technology, up to 100 megabits per second for point-to-point applications and up to 54 megabits per second for point-to-multipoint applications. Older modulation technologies are on the order of 1 megabit per second or less. Wi-Fi: Up to 7 gigabits per second for 802.11ad, 1.35 gigabits per second for 802.11ac, and 54 megabits per second for 802.11g. Licensed: Up to 800 megabits per second for licensed microwave frequency bands; however, narrowband licensed transceivers have data transmission rates on the order of 9.6 kilobits per second.
Security	◐	Varies with interference and available security capabilities. Security measures are needed to protect against man-in-the-middle attacks and unauthorized data access. Also may be subject to malicious sources of interference. Physical security should also be a consideration when selecting sites for wireless equipment.
Reliability	◐	Varies with interference and link design. A properly designed utility-owned wireless system can approach 6 nines reliability (i.e., 99.9999% uptime). However, may be subject to disruption by atmospheric anomalies, interference, and obstructions in the wireless signal path.
Distance	◐	Unlicensed: Up to 5 miles with a clear line-of-sight. Up to 1,000 feet with obstructions. Wi-Fi: Up to 0.5 miles outdoors with 802.11ac. Up to 400 feet with 802.11g. Up to 300 feet with 802.11ad. Licensed: Up to 50 miles for some point-to-point bands.
Installation Cost	○	Wireless equipment costs vary greatly depending on frequency, modulation technology, and power. Wi-Fi equipment is the least expensive because of its global popularity and widespread consumer usage. Unlicensed wireless equipment is relatively inexpensive due to relatively common usage. Licensed wireless equipment tends to be the most expensive. Installation for utility-owned wireless systems can be relatively expensive because it can include wireless path surveys, which can involve lift trucks at each end of the communications link, and the potential need for antenna towers. To save costs, antennas can be located on existing structures such as buildings, water towers, and sports arenas. Other types of signal testing, signal backhaul equipment, and amplifiers in areas with poor signal strength are other factors that can increase installation costs.
Provider Fees	●	Utility-owned wireless systems have no recurring communications provider fees, although will require utility personnel to maintain the system.
Maintenance	○	The utility is responsible for all maintenance.
Attribute Key: ● Strong ◐ Moderate ○ Weak		

Section 4: SRS Component Requirements for Communications Systems

A brief discussion is provided below for SRS components that typically require remote communications, including ESM, OWQM, and AMI. Each component's requirements with respect to communications are summarized and resources are referenced for further details.

4.1   Enhanced Security Monitoring

When an intrusion is detected at an ESM site, intrusion detection systems generate an alert signal, which consists of a relatively small amount of data that all of the communications technologies listed in Table 3-1 can transmit. To minimize response time, most ESM systems are configured to transmit intrusion alert signals as soon as they are generated, although it is possible that alert signals can be transmitted when the site is periodically polled (e.g., every 2 minutes).

ESM sites that include video monitoring can produce a significant amount of data that communications technologies with a relatively low data transmission rate (e.g., POTS), may not be capable of transmitting. Other wired technologies and digital cellular have adequate data transmission rates for video data transmission. However, transmitting a significant amount of data over a digital cellular network will require a data plan that can accommodate the amount of monthly data. Certain forms of utility-owned wireless systems may also be capable of transmitting video data. To minimize the amount of network traffic, most video monitoring systems have the option of using incident-driven video, which only transmits video data to the control center when an intrusion incident has been detected. Continuous video from remote sites is possible, but can consume available network capacity, reduce overall communications performance, and potentially lead to high usage fees. Thus, viewing live video from a remote site is typically done on an as-needed basis such as when an intrusion has been detected.

Physical security and communications systems for a typical ESM component are shown in the preliminary *architecture* diagram **(Figure 4-1)**. This diagram includes ESM equipment such as intrusion detection sensors and an access control system that processes the signals from the sensors. Video monitoring equipment is also shown, including video cameras and a video recorder. Lastly, Figure 4-1 includes equipment that is common to the OWQM component, such as an uninterruptible power supply and end-user workstations.

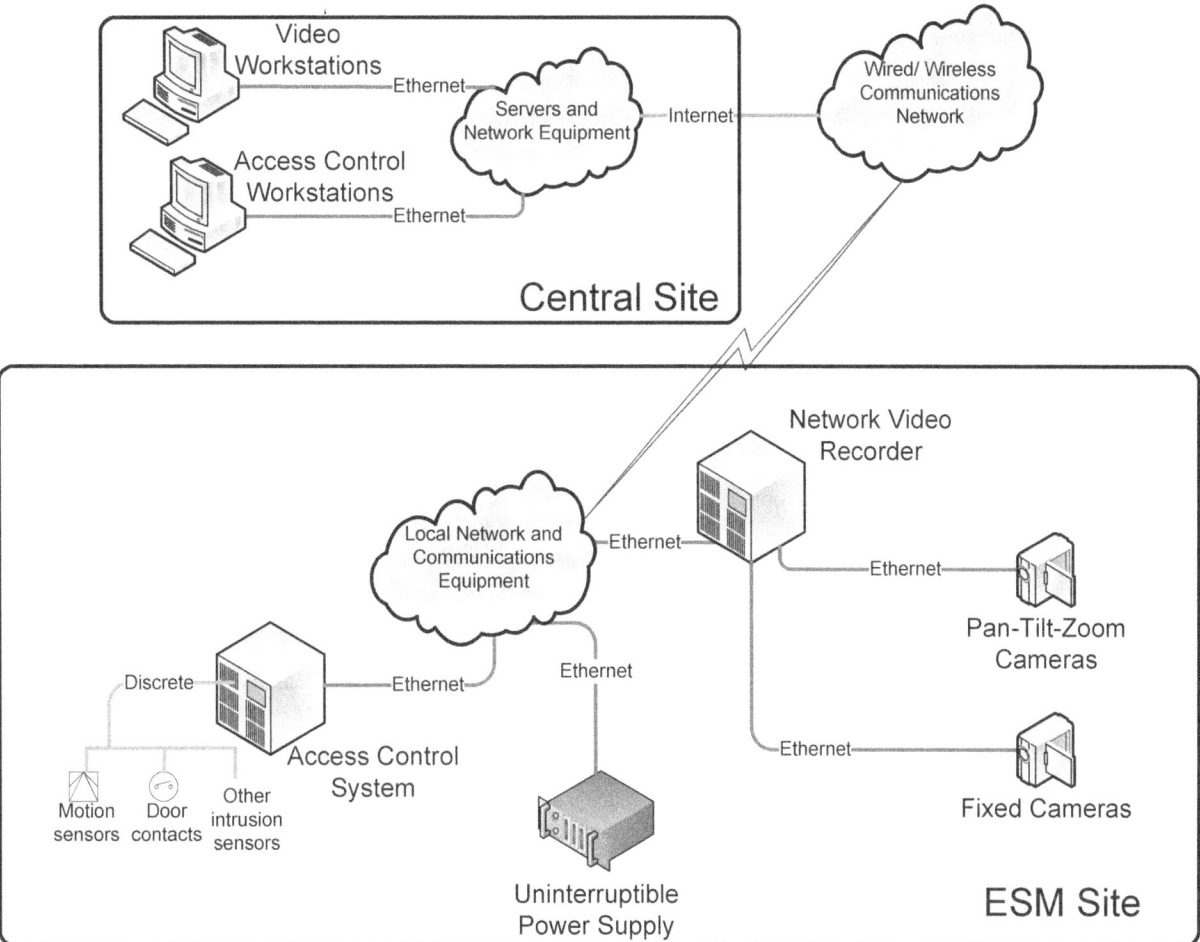

Figure 4-1. Typical Preliminary Architecture Diagram for Enhanced Security Monitoring

4.2 Online Water Quality Monitoring

OWQM involves continuous monitoring of water quality parameters, such as specific conductance, pH, and turbidity, at one or more sites in a distribution system or source waterbody. For most OWQM applications, water quality parameter values and alerts (e.g., instrument faults, power failure, communications errors) are periodically transmitted to a utility control center, usually on the order of every 2 to 5 minutes. The amount of data typically generated by an OWQM application can be accommodated by any of the communications technologies listed in Table 3-1.

However, in some cases, an OWQM site may also involve more complex parameters, such as multi-spectrum visible-ultraviolet absorbance, which can provide more information and thus detect more subtle changes in water quality. Typically, the communications technologies suitable for video transmission, as described in Section 4.1, can be considered for this type of OWQM application.

Water quality monitoring equipment and communications systems for a typical OWQM component are shown in the preliminary architecture diagram **(Figure 4-2)**. This diagram includes OWQM equipment such as a multi-parameter water quality analyzer and pressure and temperature sensors connected to a programmable logic controller for signal conversion and processing.

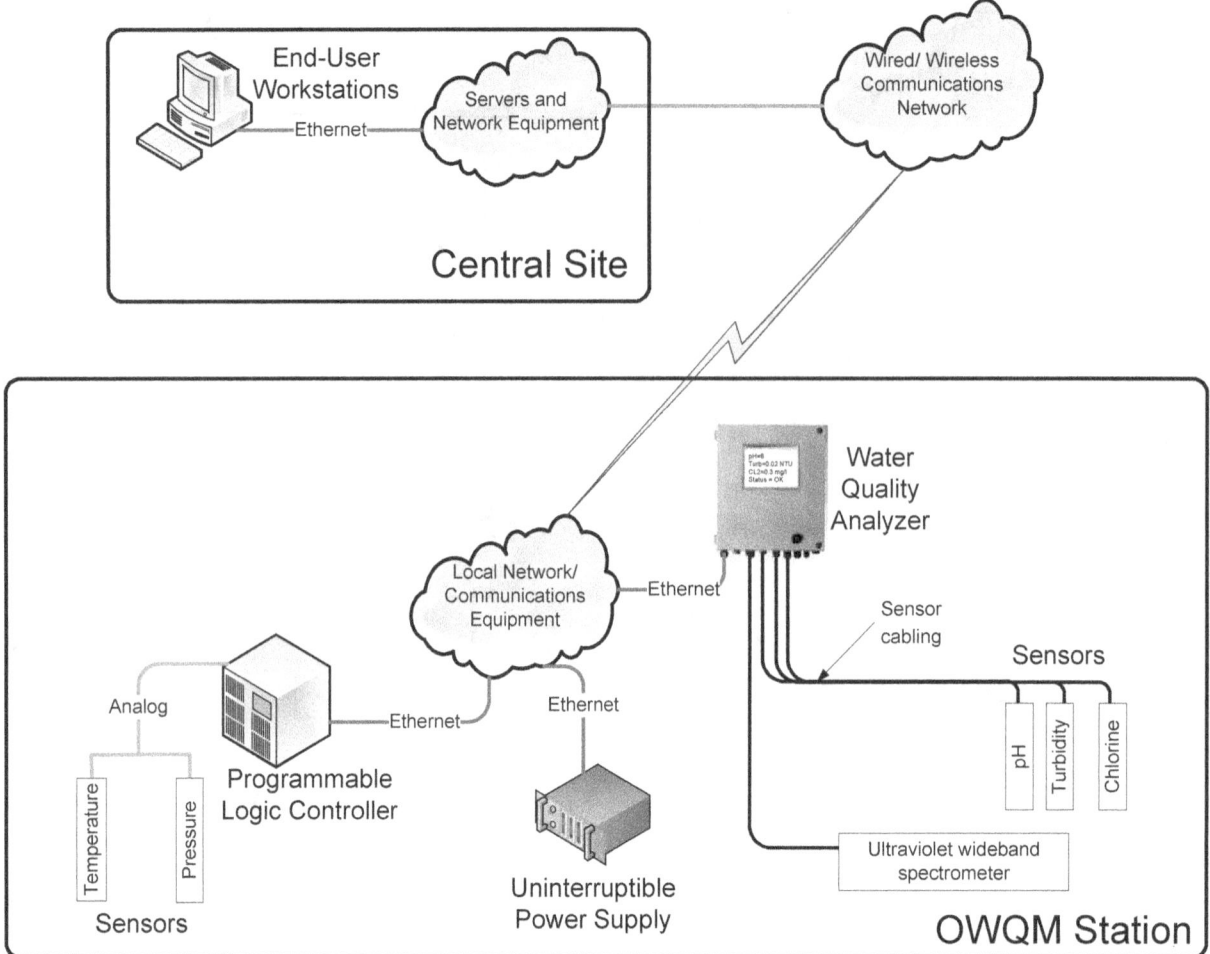

Figure 4-2. Typical Preliminary Architecture Diagram for Online Water Quality Monitoring

4.3 Automated Metering Infrastructure

Most legacy AMI systems use a proprietary wireless network consisting of one-way data transmission. Most of these proprietary networks use unlicensed frequencies, but at least one company has a licensed frequency capable of transmitting more data over longer distances. In AMI, data from each customer's water meter information unit is transmitted to data collection units located at strategic locations throughout the utility's service area. The data collection units transmit data to the AMI servers at the utility control center where the meter data management system processes the incoming metering data. **Figure 4-3** shows a typical legacy AMI communications system.

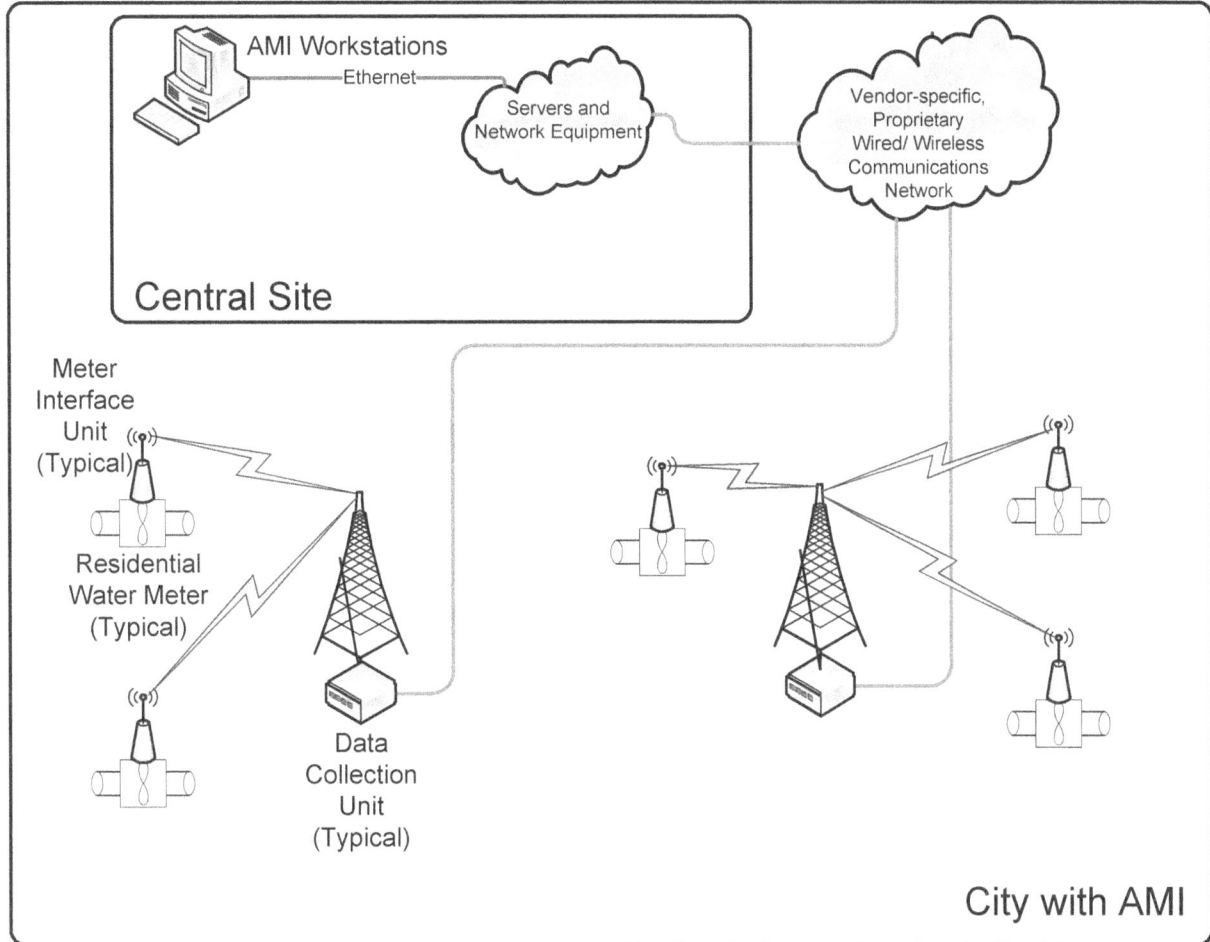

Figure 4-3. Typical Legacy Automated Metering Infrastructure Communications System

An AMI system may also include leak detection sensors, district flow meters, and fire hydrant-based pressure sensors at critical locations in a distribution system. Generally, the meters take hourly readings and transmit the accumulated data every four to six hours. The current state of battery technology is such that more frequent wireless transmissions would result in unacceptable battery life. However, advances are being made in battery and power-harvesting technology such that more frequent data sampling and wireless transmissions from customers' meters can occur while maintaining an acceptable battery life expectancy.

Currently, most AMI companies are migrating away from proprietary *one-way communications* systems toward cellular communications. Furthermore, *two-way communication* is an option with cellular technologies, which can be used to implement remote shut-offs, an important function with operational and SRS significance. Specifically, remote shut-offs allow service to be discontinued without having to dispatch personnel to the residence. Furthermore, newer meters that are equipped with tampering and backflow sensors may be used in conjunction with remote shut-offs to prevent contaminants from being injected into the distribution system when a tamper or backflow incident is detected.

4.4 Component Summary

Table 4-1 lists the typical types of technologies to consider for ESM, OWQM, and AMI applications. Digital cellular technology has reached a point where it is suitable for data-

> **HELPFUL HINT**
>
> A utility with a limited budget that is not implementing video monitoring as part of its ESM component might consider using a third-party alarm service, which can use cellular technology or POTS for communications.

intensive applications such as ESM video and complex parameter data from OWQM. Wired technologies, except POTS, are also suitable for all SRS applications, although availability at remote locations may be limited. Utility-owned wireless systems are generally better suited for applications that do not have video or complex parameter data, although newer wireless technologies have greatly increased in data transmission rate.

Table 4-1. Typical Communications Technologies for SRS Components

Component Type	Digital Cellular	Utility-owned Wireless	Wired Technologies
ESM without video (alerts-only)	●	●	●
ESM with video	●	○1	●2
OWQM non-spectral	●	●	●
OWQM spectral	●	○1	●2
AMI	●	●3	---
Attribute Key: ● **suitable** ○ **may not be suitable**			

Notes:

1. Suitability will depend on type of utility-owned wireless system used. See Section 3.2.2 for more information on utility-owned wireless systems.
2. Except for POTS.
3. Some AMI systems communicate via wireless equipment that uses the AMI vendor's proprietary protocol. The AMI wireless equipment can be owned by the utility or leased from the AMI vendor.

Section 5: Selecting a Communications System

5.1 Identifying Communications Options

The evaluation criteria developed in Section 2 should be presented to and discussed with a utility's internal data communications department, city or state-owned metropolitan communications network providers, and third-party providers (e.g., the local telephone company, cellular carriers, and cable television companies). For providers that have an existing relationship with the utility, such as a cellular contract for smart phones, the utility can consult its account representative to leverage the current agreement and establish cellular service at SRS locations. For providers that do not have an existing relationship with the utility, the department to contact for information on communications can vary and may be called any of the following terms:

- Internet of Things
- Machine-to-Machine
- ***Smart Cities***
- Business-class

It may be possible that no communication provider in your area has service available at all of the sites and that multiple providers or utility-owned systems may be needed. Multiple providers or a combination of third-party provided and utility-owned systems may also be preferred for redundancy, competitive pricing, or other utility-specific needs.

A communications system should be designed to serve multiple SRS components, which can result in cost savings. However, the feasibility of a shared system will depend on each component's data communications requirements and the costs associated with the available communications options. Thus, separate communications systems—each dedicated to a component—can also be considered, but may be less cost effective and require more maintenance by the utility. Coordination among components is strongly recommended to optimize utilization of shared communications systems.

> **CASE STUDY**
>
> A large utility had an communications system that included a combination of existing utility-owned fiber optic connections and 4G cellular to areas without existing fiber optic infrastructure. A mixture of technologies is common for SRS applications.

5.2 Alternatives Evaluation and Selection

After developing a list of the available communications technologies, the utility can perform a high-level screening to eliminate any alternatives that clearly would not meet the evaluation criteria developed in Section 2, especially budget constraints. The remaining alternatives can then be evaluated using the process described in *Framework for Comparing Alternatives for Water Quality Surveillance and Response Systems* based on the evaluation criteria. The selected technology (or technologies) will then be used to develop a preliminary design and detailed design, which will be implemented as described in Section 6.

Section 6: Design and Implementation

If a utility determines that it will need to implement a new communications system to support its SRS, it will need to design and implement the system in a manner that meets its established evaluation criteria. The design and implementation process can be divided into the steps described below; a detailed discussion of each step is provided in the sections that follow.

1. **Document System Requirements:** Determine the location, necessary SRS equipment, and data generation rates at each site.
2. **Review with Provider:** Review the preliminary design with the selected communications provider.
3. **Conduct Field Assessment:** Visit each site to determine existing wired communications infrastructure, wireless signal strength, and constraints.
4. **Develop Detailed Design:** Refine the preliminary design based on the results of Steps 2 and 3.
5. **Initiate Implementation:** Install the equipment and commission each remote site to ensure that the data communications requirements are met.

As a utility works through the above process, the information received from the provider in Step 2 or obtained from the field assessment in Step 3 may require the utility to revisit the previous step or revise the preliminary design. An iterative approach is common especially when working with complex requirements and a limited budget.

6.1   Document System Requirements

Transmitting data from the SRS equipment to the user at the control room requires a variety of intermediate communications devices, and their arrangement can vary depending on utility policies, security requirements, and equipment needs. Thus, an important first step in the design process is to document the requirements for the communications system. A preliminary architecture diagram for each site, such as those shown in Section 4, can be used to visualize the overall arrangement of and connections between network devices and SRS equipment (e.g., video cameras, door sensors, motion sensors, water quality analyzers). Also, the utility should calculate the data generation rates and data generated per month at each site, and determine the physical location of each site (street address or Global Positioning System coordinates). Location information will be needed by the communications providers to determine if an SRS site is within their service areas.

The location and number of workstations required for SRS interfaces should be approximated when developing the architecture. Although the number of workstations does not necessarily impact the communications system design, it is an important consideration when developing the *information management system*. Information management system design is beyond the scope of this document, but is described in more detail in *Guidance for Developing Integrated Water Quality Surveillance and Response Systems* and the *Dashboard Design Guidance for Water Quality Surveillance and Response Systems*.

> **HELPFUL HINT**
>
> Developing an architecture diagram for a communications system, new or existing, is essential for documenting the data paths, SRS equipment, and networking devices. Without one, a system can be challenging to manage, maintain, and secure.

6.2   Review with Provider

The requirements for the communications system should be reviewed by and discussed with the communications provider or internal utility communications department. The provider can provide technical details such as the equipment to be provided in the clouded portions of the preliminary architecture and security practices that can be implemented. The provider can also provide input regarding schedule and costs for equipment and installation.

6.3   Conduct Field Assessment

For utility-owned wireless communications systems, a signal path survey, as described in Section 3.2.2, may be required. This effort may need a separate contractor that specializes in signal path surveys, or can be done by utility personnel if the appropriate equipment is available (a bucket truck, antennas, antenna cabling, test equipment, etc.). A spectrum analysis should also be performed at each site to identify potential sources of interference. If the utility is implementing a wireless system that requires broad areas of coverage, software is available for conducting a desktop-level, system-wide propagation survey that can identify dead spots.

For digital cellular systems, site surveys are relatively straightforward and typically consist of visiting the SRS site with a smartphone provided by the carrier and viewing the Received Signal Strength Indicator value. Check with the carrier regarding the minimal level of Received Signal Strength Indicator that will provide acceptable performance.

For wired systems, an on-site inspection may be needed to determine the existing equipment and services currently available at the facility.

6.4   Develop Detailed Design

The detailed design should be based on the requirements, the communications provider's technical information, and field assessment results. The detailed design should include equipment manufacturer and models and the connections between devices. For wireless technologies, antenna heights and locations should be specified. If the wireless solution is utility-owned, the locations of repeater stations may also be needed where line-of-sight is not available or the distance between a remote site and the control center exceeds the transmitter/receiver range. Security, *data directionality,* data transmission rate limits, redundancy, network topology, and data usage are also detailed design considerations, each of which are described below.

6.4.1 Security

Security should be considered at this stage, and industry-standard best practices such as strong password requirements, antivirus protection, VPN connections, and traffic segregation using firewalls are strongly recommended for all communications connections, especially those that are exposed to the Internet.

VPNs provide private encrypted links, often referred to as "tunnels," between two points to protect against cyber threats intent on reading or injecting data into an existing connection, often referred to as a "man-in-the-middle" attack.

Firewalls are recommended for regulating the traffic between a utility-owned network and external networks by blocking traffic that does not satisfy user-specified rules. Firewall rules can specify, for example, that traffic must only access certain ports, be generated by a pre-approved application, or be solicited from a user's application. Furthermore, some firewalls can be programmed to detect and block suspicious data packets. Data can also be encrypted using technologies such as Secure Socket Layer or Internet Protocol (IP) Security as an added means of ensuring data integrity and security. As cyber threats to utilities become more sophisticated and prevalent, it is strongly recommended that utilities work with *cybersecurity* experts to keep their networks and computing policies, procedures, and protective measures up to date. Consult the *Framework for Improving Critical Infrastructure Cybersecurity* for more information on cybersecurity measures.

6.4.2 Data Directionality

Whether data will be transmitted only from the remote site to the control center (one-way) or also be transmitted from the control center to the remote site (two-way) is an important distinction for the communications system design. Examples of two-way data flow include the positioning of pan-tilt-zoom cameras and initiating sample collection at remote ESM and OWQM facilities by personnel located at the control center. All technologies are technically capable of two-way data flow, although the equipment may need to be configured to enable this function. Furthermore, some utilities may require additional cybersecurity countermeasures for control signals transmitted to a remote site.

6.4.3 Data Transmission Rate Limits

When using a communications link to transmit multiple SRS datastreams such as ESM video and OWQM spectral data, imposing data transmission rate limits on each datastream may be needed to prevent one datastream from impacting the performance of the others. When selecting network equipment, confirm that this function is included.

6.4.4 Redundancy

A utility might consider a redundant communications system for critical sites. For example, a wireless technology, such as digital cellular, could be used to provide backup communications capability at a site connected to the utility's fiber optic network, especially if the fiber optic cabling passes through an area undergoing extensive construction and excavation. The utility can perform a *benefit-cost analysis* to weigh the likelihood and consequences of communications loss from a site versus the cost of the additional provider fees or utility-owned infrastructure.

> **CASE STUDY**
>
> A utility used a fiber-optic connection to transmit ESM video and real-time process control and monitoring data (i.e., SCADA data). Without imposing limits on either datastream, the ESM video consumed all of the available transmission capacity and SCADA data performance was significantly impacted. The utility reconfigured its network equipment to impose limits on the video datastream, and an acceptable level of performance for ESM video and SCADA data was achieved.

6.4.5 Topology

Redundancy can also be achieved via certain types of network topology, which is the arrangement of allowable data paths between SRS sites. Network topology should incorporate data directionality requirements when designing a utility-owned communications system to provide an acceptable level of performance and redundancy. For a third-party communications system, the network topology is the provider's responsibility and is transparent to the utility. Three commonly used types of network topologies for SRS applications are star, tree, and mesh. Each are further described below.

Star Topology

The star topology consists of a central site to which all remote sites communicate. A benefit of the star topology is that each remote site is directly connected to the central site without any intermediaries that could delay data transmission. However, no redundant pathways of communication exist if a link fails. Furthermore, the central site is a single point of failure, so extra precautions are usually taken to ensure the central site's uptime, such as implementing redundant servers and communications equipment, and housing the equipment in a physically secure location. This topology is commonly used for SRS applications with remote sites that need to communicate only with the central site—not each other—such as ESM and OWQM. See **Figure 6-1** for a diagram of a star topology.

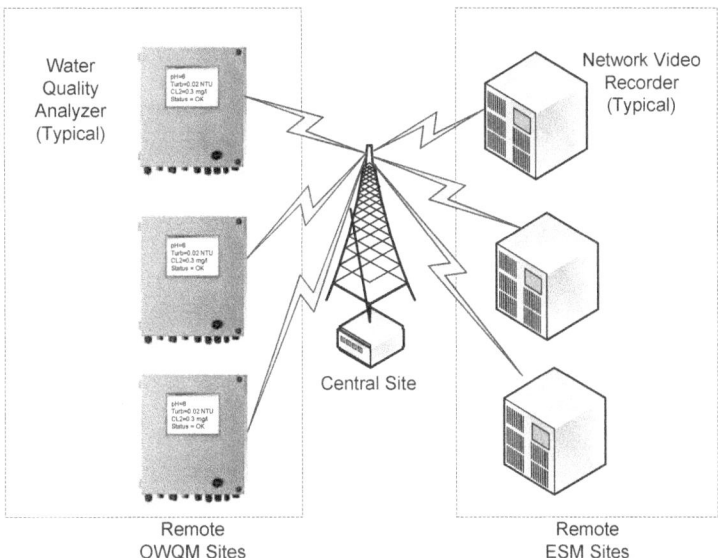

Figure 6-1. Star Topology

Tree Topology

The tree topology is a hierarchical arrangement of data paths where data originates at the bottom layer of the tree and works its way to the top of the tree via multiple layers. The benefit of this topology is that it allows the remote sites to use a communications technology with limited range to minimize power consumption, which is critical for battery-powered devices such as AMI meter interface units. Thus, AMI vendors prefer the tree topology for their communications networks. This topology requires that data collection sites are strategically located throughout the utility's service area and overlapping coverage areas of data collectors can provide some redundancy. For example, if one data collector is down, some of its remote sites may be able to reroute their data transmissions to a neighboring data collector. See **Figure 6-2** for a tree topology diagram.

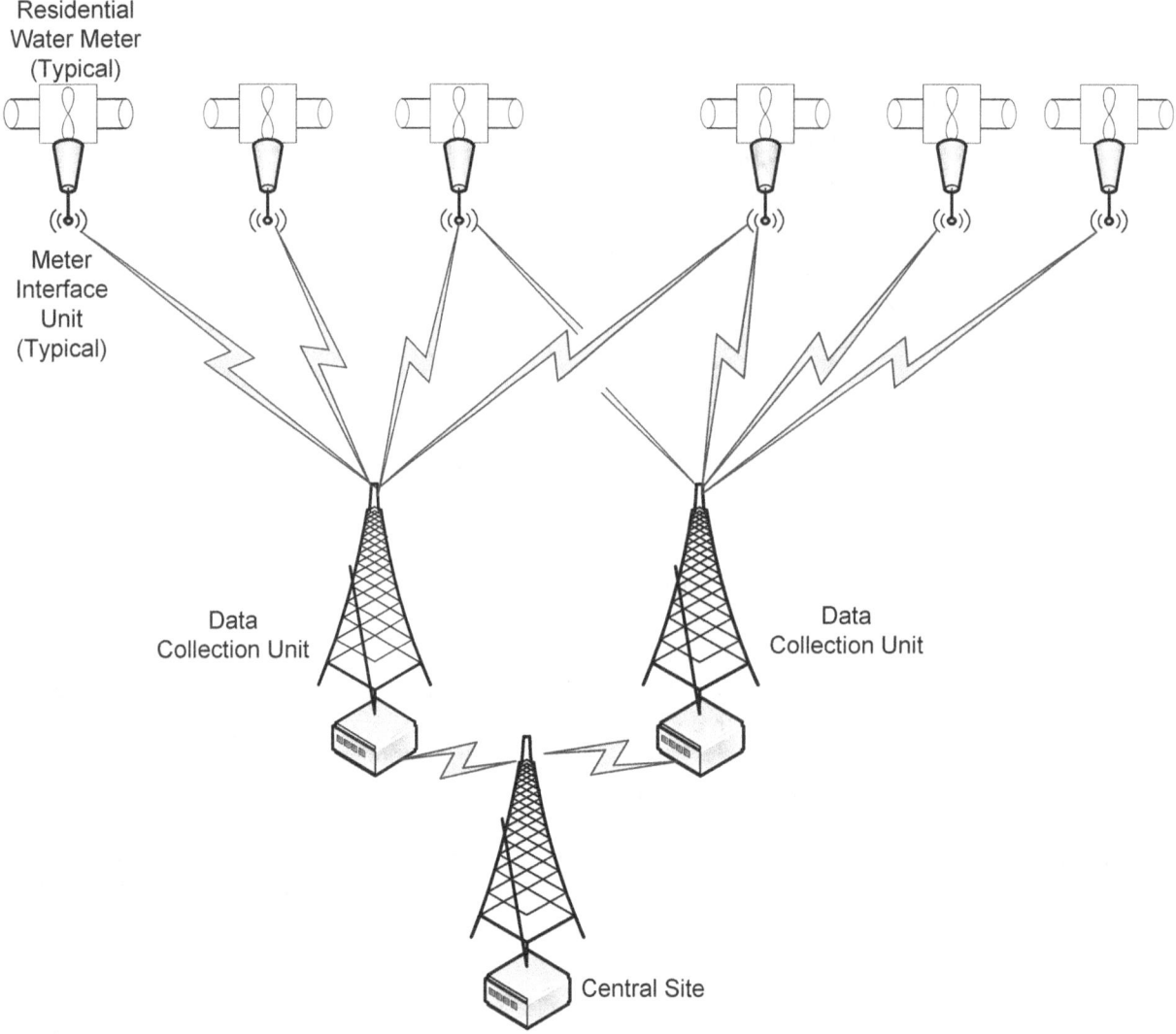

Residential
Water Meter
(Typical)

Meter
Interface
Unit
(Typical)

Data
Collection Unit

Data
Collection Unit

Central Site

Figure 6-2. Tree Topology

<u>Mesh Topology</u>

The mesh topology is an arrangement of data paths where each site acts as a repeater for the network and cooperates with other sites to distribute data throughout the network. The advantage to this topology is its redundancy. If one site or more sites have failed, the remaining sites can still communicate by routing their messages around the failed sites. However, there will be a delay in transmission when data is retransmitted through an intermediate site. **Figure 6-3** shows a mesh topology that is a fully connected network, where every site is capable of communicating with every other site in the network.

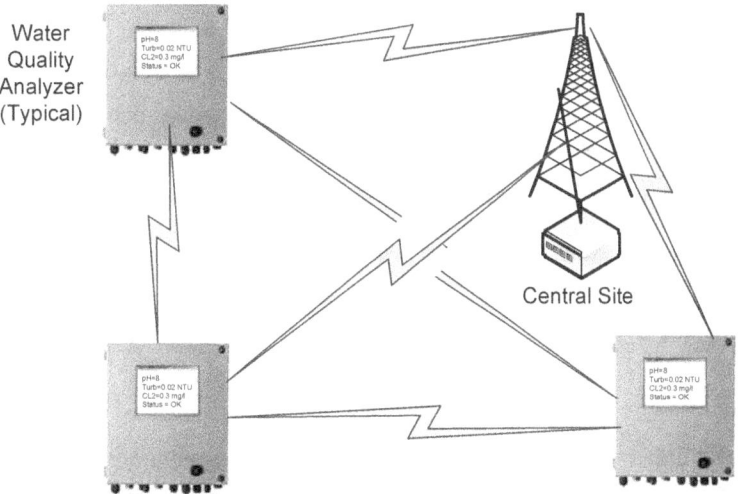

Figure 6-3. Mesh Topology – Fully Connected

In practice, a fully connected mesh topology is usually not feasible because of distance limitations between sites, so a partially connected mesh topology is implemented. In a partially connected mesh arrangement, each site can communicate with nearby sites (rather than all other sites) for an acceptable level of redundancy. A partially connected mesh topology, which could be used for an SRS application, is shown in **Figure 6-4**.

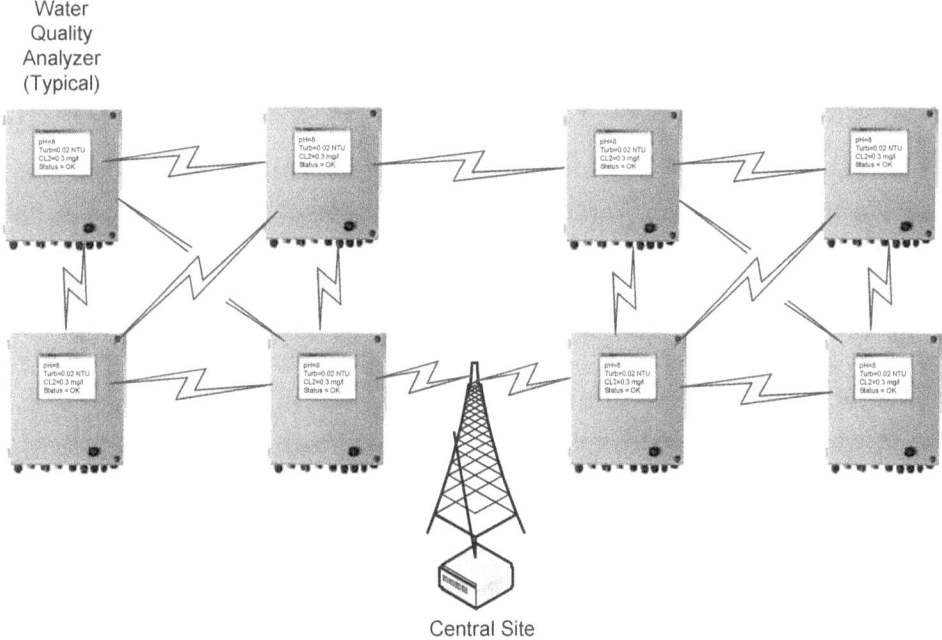

Figure 6-4. Mesh Topology – Partially Connected

LESSON LEARNED

A large utility configured its ESM system such that video data from a remote site was transmitted to all other remote sites in addition to the control center for maximum flexibility (i.e., a mesh topology). However, this resulted in poor network performance, and the utility revised its ESM configuration such that video data from a remote site was transmitted only to the control center (i.e., a star topology).

This example shows that a mesh topology may experience performance issues with high data generation rate systems such as video monitoring. Conversely, with lower data generation rate systems, mesh topologies can provide effective communications with a significant amount of redundancy. Thus, a utility should carefully consider the operational benefit and performance implications of its network topology when developing a preliminary design.

6.4.6 Data Usage

Data usage is the amount of data transmitted to or from each SRS site over a given period, typically a month. This metric should be estimated for each SRS site if using a technology that bills on a data usage basis, such as digital cellular.

Digital cellular data plans are tiered in gigabyte-per-month increments, and users are subject to additional fees when the amounts of data generated during a month exceed their allotments. Some cellular providers offer shared data plans that allow multiple devices to draw from a common allotment of data per month. The utility may want to start with a conservatively high estimate of data per month and track its usage to determine whether a lesser data plan can be used. After systems are properly commissioned, less data usage is expected.

This metric can be estimated empirically with a pilot-scale evaluation similar to the process for determining the data generation rate for new equipment (discussed in Section 2.2.1) and using data records for existing devices. If pilot testing is not preferred or not possible, this metric can be estimated using the calculations described in the Appendix.

6.4.7 Detailed Communications Architecture Diagrams

After obtaining technical specifications from the provider, a detailed *communications architecture* diagram can be developed. The equipment and communications clouds shown in the preliminary drawings can be completed with the specified hardware and technologies. For example, in **Figures 6-5** and **6-6**, the "servers and network equipment" and "wired/wireless communications network" clouds have been replaced with specific servers and a digital cellular service.

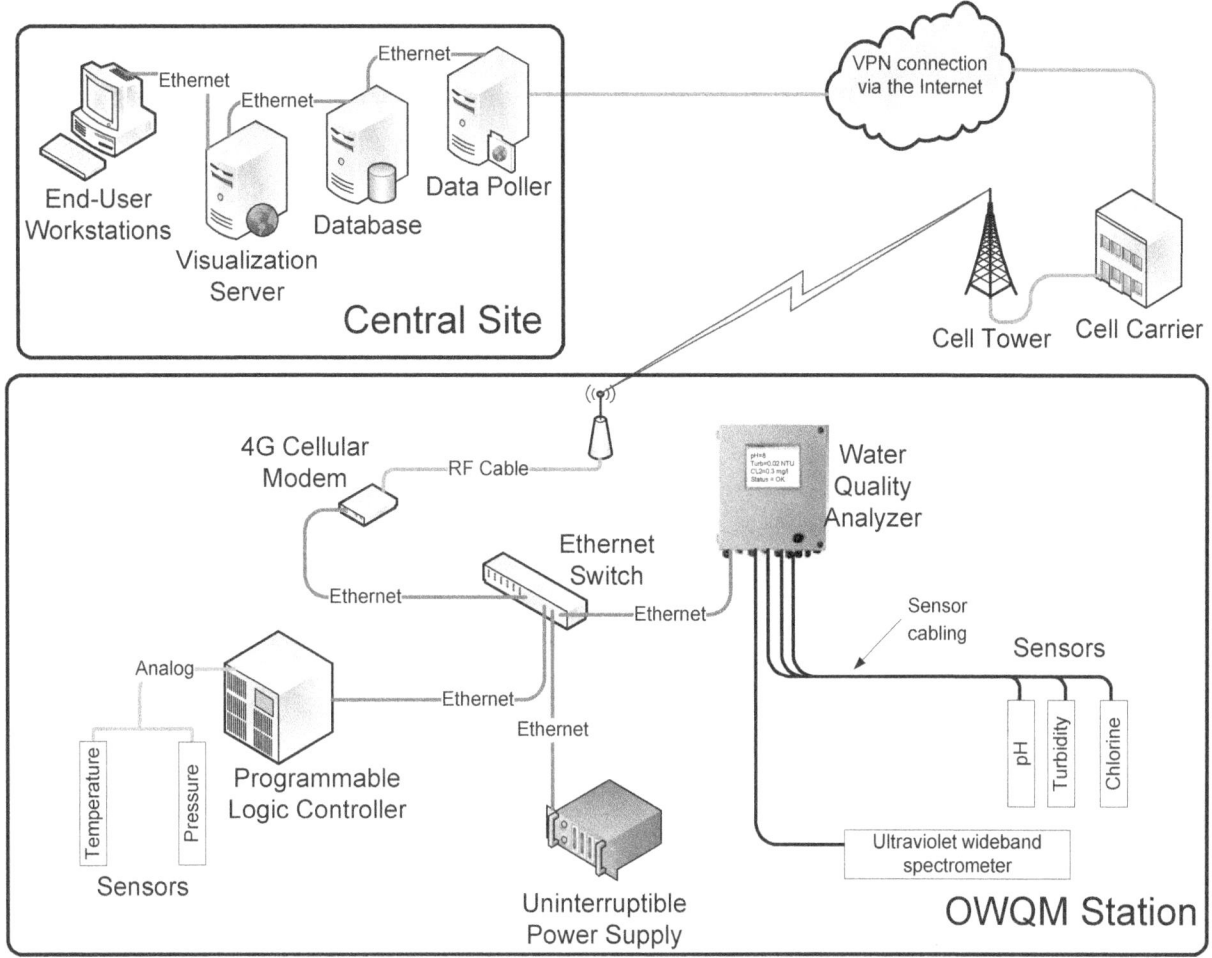

Figure 6-5. Typical Detailed Communications System Architecture for Online Water Quality Monitoring

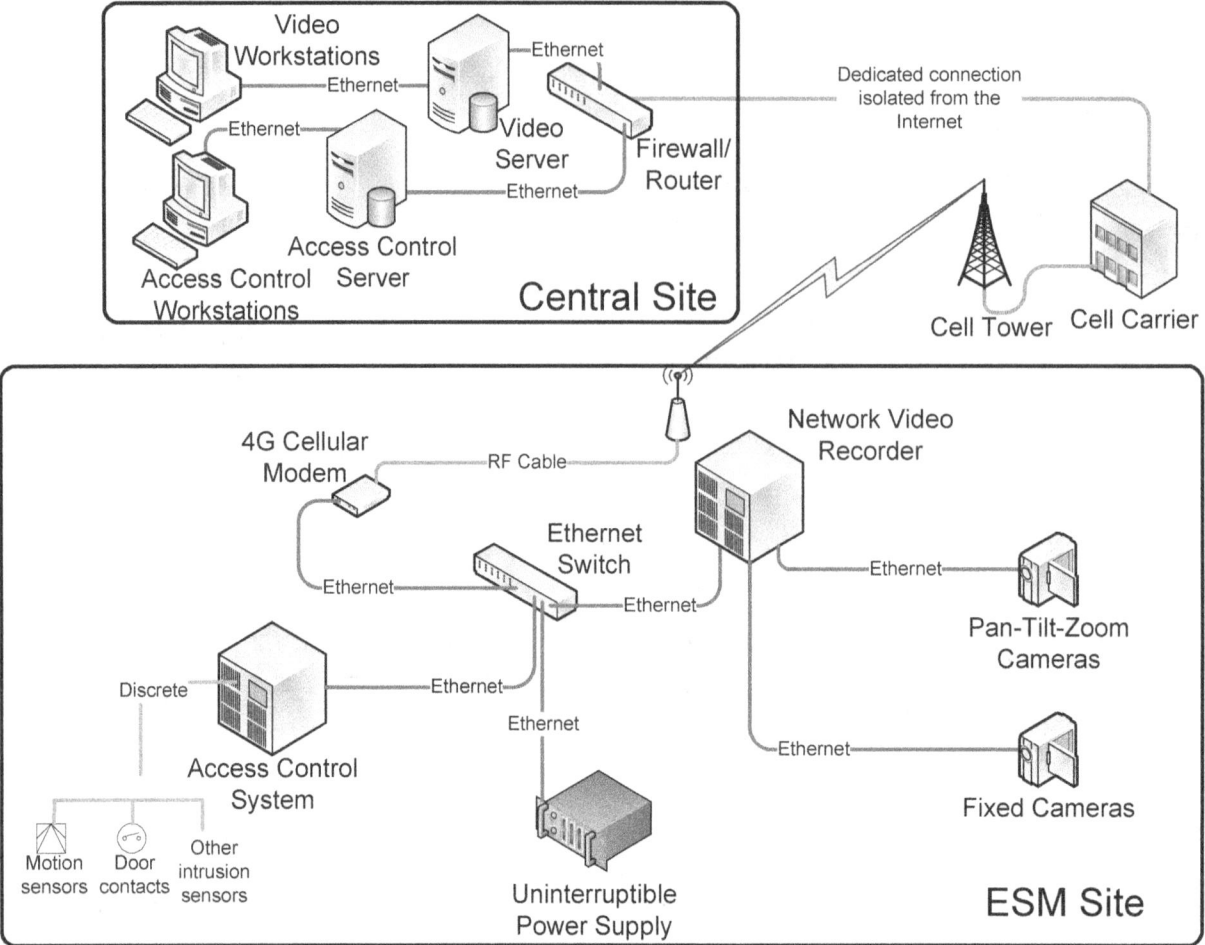

Figure 6-6. Typical Detailed Communications System Architecture for Enhanced Security Monitoring

For a comprehensive view of the overall communications system, the utility may also consider consolidating the detailed architectures from the SRS components into a single diagram. Alternately, if multiple communications technologies are being used, a detailed diagram may be developed for each technology. Consider an example application where ESM and OWQM sites will use a combination of digital cellular and fiber optic connections. In this case, one detailed diagram could be developed that shows all of the ESM and OWQM digital cellular sites, and another detailed diagram could show the fiber optic connections.

6.4.8 Contract Documents

For a utility-owned system, using the information developed for the detailed design architecture for each component, contract drawings and specifications can be developed and issued to be bid upon by contractors. The contract documents must include strict requirements for a robust submittal and review process and a thorough and transparent commissioning process. It may also be beneficial to include requirements for the contractor to provide a certain number of hours of post-installation support and associated travel costs to address issues that may occur after system commissioning.

If the utility has the option, soliciting only preapproved vendors (e.g., through a statewide contract for communications services and equipment) is strongly recommended (Mix et al., 2011). This can lead to a shorter bidding process by avoiding the lengthy public advertisement and bidding stages and lower costs associated with a negotiated statewide pricing agreement. Furthermore, bid documents would not be released to the general public, keeping sensitive content more secure than with public bids.

For a system with third-party provided communications, the amount of publically-bid work may be minimal. However, most communications providers will require a service-level agreement to establish costs, uptime, reliability, and address outages. Negotiating the terms of this agreement with the provider is an important step that can ensure that the utility's needs are met within its available budget. See Section 2.4 for more information on service-level agreements.

6.5   Initiate Implementation

After a contractor is successfully procured, construction shop drawings and submittals should be carefully reviewed by the utility's personnel and design engineer to ensure conformance with the contract documents. Implementation should not commence until the shop drawings and submittals have been approved by reviewers. During implementation, the utility should have a qualified inspector on site to ensure that installation and commissioning are performed as shown on the shop drawings and submittals, and conform to the contract documents. For more information on commissioning, consult *Commissioning Security Equipment – Getting it Right the First Time*.

After implementation, a utility should consider meeting annually to evaluate the performance of their communications network and discuss changes to utility data requirements and technology upgrades. During this meeting, competitive alternatives may be discussed to take advantage of newer technologies that cost less while phasing out older technologies that are becoming obsolete. Periodic meetings with the communications provider or internal utility communications department may also be held to discuss service-related issues and future plans.

Section 7: Looking Ahead

Traditionally, communications systems consist of remote sites communicating with a central facility, where all decision-making occurs. This communications architecture is called interconnect, and typically uses a star or tree topology as shown in Figures 6-1 and 6-2. However, communications technologies and computer processors have become cheaper, faster, smaller and consume less power. Innovation has led to two important changes in communications architecture:

1. **Peer to peer (P2P) Connectivity:** In this communications architecture, most of the data communications occurs among remote sites with minimal data transmitted to the central facility on an as-needed basis. This communications architecture typically uses a mesh topology as shown in Figures 6-3 and 6-4.

2. **At the Edge Processing:** Data processors at the remote sites analyze data gathered locally and received from nearby sites and perform decision-making functions, without or with minimal direction from the central facility.

A critical benefit of these two changes is that the amount of data to be transmitted to the central facility is significantly reduced, which decreases transmission costs and improves network performance. Furthermore, by de-emphasizing the role of the central facility, a potential bottleneck or point of failure is minimized. Thus, a utility that uses this type of communications architecture can add significant amounts of remote sites to their communications systems without overtaxing the communications links to their central facility.

Figure 7-1 shows an example of a traditional OWQM application using an interconnect communications architecture. A water quality analyzer in the distribution system periodically transmits data values every 5 minutes to an *anomaly detection system (ADS)* running on a server at the utility control center. The processing of these large data sets may be distributed to occur on multiple machines. For this example, data values include: pH, turbidity, chlorine, oxidation reduction potential (ORP), and total organic carbon (TOC). If the ADS detects contamination, coded logic allows the ADS to transmit a close command signal to the appropriate automated valve in the distribution system to prevent the spread of possible contamination.

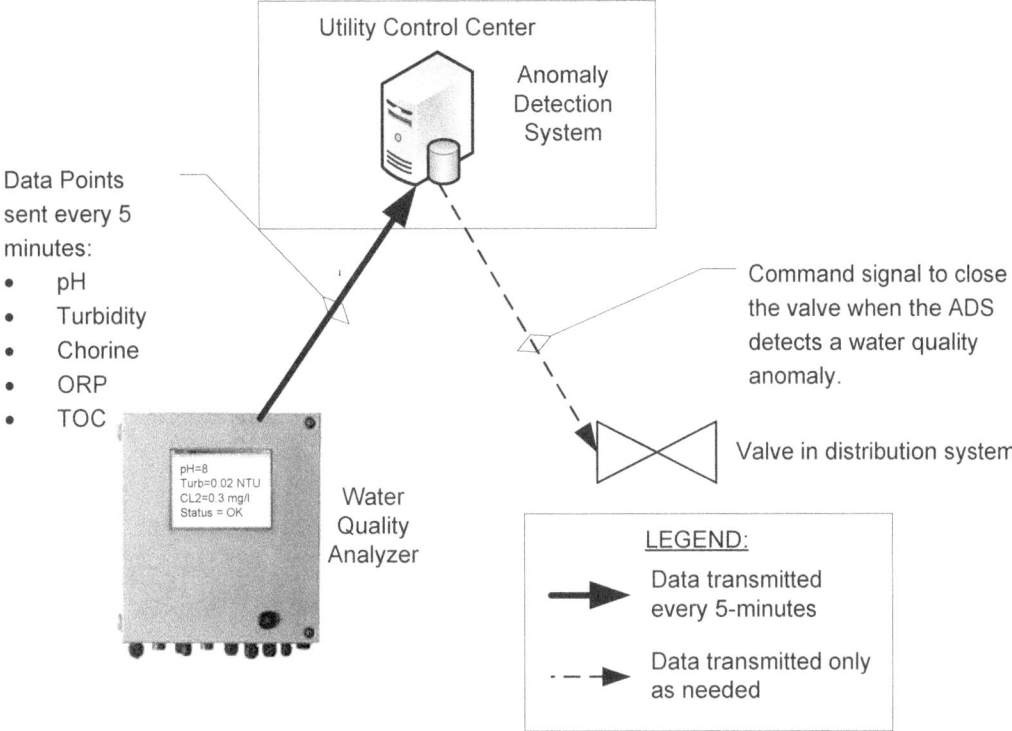

Figure 7-1. Example OWQM Application with Traditional Interconnect Architecture

A common OWQM configuration consists of 10 to 20 monitoring stations measuring 5 parameters and polling data every five minutes. Most communications systems using traditional interconnect architecture can handle traffic from such a configuration. However, with the potential for water quality sensors to be included in district and residential meters, the amount of data points transmitted back to the control center can exceed the capacity of an interconnect architecture and drive up communications costs, especially at 5-minute intervals. As the amount of monitoring stations in the distribution system grows, the need for P2P and at-the-edge processing becomes more pressing.

Figure 7-2 shows the same OWQM application from Figure 7-1 using a P2P and at-the-edge processing. The water quality analyzer runs the ADS locally, so data is only transmitted to the control center when requested by utility personnel. When the ADS detects anomalous conditions, the water quality analyzer transmits a close command signal to the valve. At-the-edge processing is essential for this OWQM application, especially with a large number of monitoring stations.

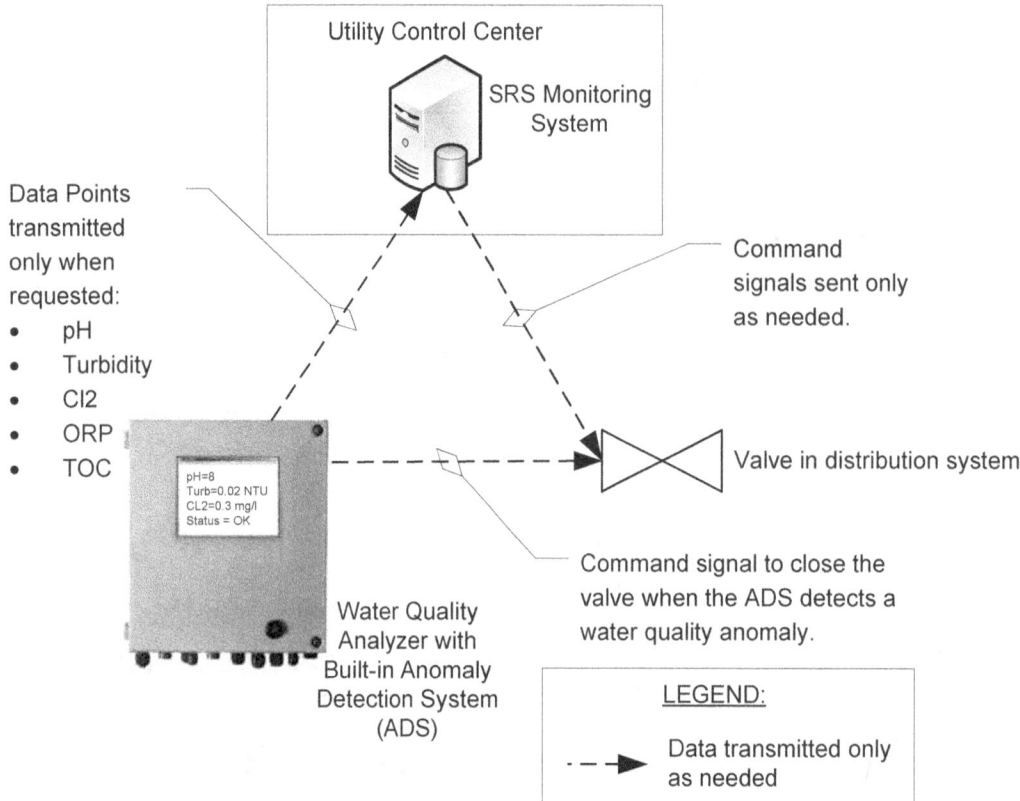

Figure 7-2. Example OWQM Application with P2P Architecture with At-The-Edge Processing

The examples shown in Figures 7-1 and 7-2 show a fully automated OWQM/ADS response system for the purpose of highlighting the differences between communications architectures. However, for most utilities, the decision to implement operational changes such as closing a valve or shutting off a pump would require human intervention and decision-making.

Resources

Section 5.2: Alternatives Evaluation and Selection

Framework for Comparing Alternatives for Water Quality Surveillance and Response Systems (EPA, 2015)

http://www.epa.gov/sites/production/files/2015-07/documents/framework_for_comparing_alternatives_for_water_quality_surveillance_and_response_systems.pdf

This document provides guidance for selecting the most appropriate design from a set of viable alternatives. It guides the user through an objective, stepwise analysis for ranking multiple alternatives and describes, in general terms, the types of information necessary to compare the alternatives. EPA 817-B-15-003, June 2015.

Section 6.1: Preliminary Design

Guidance for Developing Integrated Water Quality Surveillance and Response Systems (EPA, 2015)

http://www.epa.gov/sites/production/files/2015-12/documents/guidance_for_developing_integrated_wq_srss_110415.pdf

This document provides guidance for applying system engineering principles to the design and implementation of a Water Quality Surveillance and Response System to ensure that the SRS functions as an integrated whole and is designed to effectively perform its intended function. Section 4 provides guidance on developing information management system requirements, selecting an information management system, and IT master planning. EPA 817-B-15-006, October 2015.

Dashboard Design Guidance for Water Quality Surveillance and Response Systems (EPA, 2015)

http://www.epa.gov/sites/production/files/2015-12/documents/srs_dashboard_guidance_112015.pdf

A *dashboard* is a visually-oriented user interface that integrates data from multiple Water Quality Surveillance and Response System components to provide a holistic view of distribution system water quality. This document provides information about useful features and functions that can be incorporated into an SRS dashboard. It also provides example user interface designs. EPA 817-B-15-007, November 2015.

Section 6.4.1: Cybersecurity

Framework for Improving Critical Infrastructure Cybersecurity (NIST, 2014)

http://www.nist.gov/cyberframework/upload/cybersecurity-framework-021214-final.pdf

Uses a common language to address and manage cybersecurity risk in a cost-effective way based on business needs without placing additional regulatory requirements on businesses. The framework focuses on using business drivers to guide cybersecurity activities and considering cybersecurity risks as part of an organization's risk management processes. The Framework consists of three parts: the Core, the Profile, and the Implementation Tiers.

Section 6.5: Installation

Commissioning Security Equipment – Getting it Right the First Time (EPA, 2012)

https://www.epa.gov/sites/production/files/2015-06/documents/commissioning_security_systems_for_drinking_water_utilities.pdf

Discusses commissioning of security systems and provides a step-wise commissioning process and forms for use by drinking water utilities. The objectives of commissioning are to ensure that systems perform as designed and meet the owner's needs. Although this document focuses on security equipment and reducing nuisance alerts, its nine-step approach can also be applied to a communications system.

Section A.3.1: Approximate the Maximum Data Generation Rate for Each Remote Site

Online Water Quality Monitoring Design Guidance For Distribution System Monitoring (EPA, 2016)
https://www.epa.gov/waterqualitysurveillance
Describes the OWQM component and its design elements, including Data Generation, Event Detection, and Alert Investigations.

References

Menon, A. (2015). Merging Smart Cities with Smart Water. *Proceedings of the Smart Water Network (SWAN) Conference*. London, England.

Mix, N., Golembeski, J., Baranowski, C. (2011). Understanding Information Security Terminology and Governance for Drinking Water and Waste Water Utilities. *Proceedings of the Water Security Congress*. Nashville, Tennessee. (Manuscript for Abstract ID 35294)

Glossary

alert. An indication from an SRS surveillance component that an anomaly has been detected in a datastream monitored by that component. Alerts may be visual or audible, and may initiate automatic notifications such as pager, text, or email messages.

anomaly detection system (ADS). A data analysis tool designed to detect deviations from an established baseline. An ADS may take a variety of forms, ranging from complex computer algorithms to thresholds.

architecture. The fundamental organization of a system embodied in its components, their relationships to each other, and to the environment, and the principles guiding its design and evolution. The architecture of an information management system is conceptualized as three tiers: source data systems, analytical infrastructure, and presentation.

Automated Metering Infrastructure (AMI): A potential surveillance component of an SRS. AMI consists of battery powered wireless device at each water meter that transmits water usage data to a utility's control center for processing. Although AMI data is primarily used for customer billing, this data can also be used to detect customer leaks and reverse flow conditions. For operational monitoring, an AMI can include leak detection, pressure, and temperature sensors at strategic points in a distribution system. Furthermore, newer AMI systems have the ability to include a remotely controlled valve, a tamper sensor, and water quality sensors at the meter, which have significant SRS potential.

benefit-cost analysis. An evaluation of the benefits and costs of a project or program, such as an SRS, to assess whether the investment is justifiable considering both financial and qualitative factors.

"best effort" service. A level of maintenance for a third-party provided communications service where the provider is not obligated to maintain the subscription service level or pay penalties whenever the actual service level falls below the subscription service level. Additionally, the provider's responsiveness for repairing outages will vary with the amount of higher priority issues that the provider is addressing.

commissioning. The process of testing a newly-installed or modified system for proper operation, configuration, and calibration.

communications architecture. The arrangement and connectivity between network devices used to transmit data from one point to another.

communications provider. An entity that allows customers to use its network to transmit data. This is includes third-party service providers such as cellular or telecommunications carriers and cable television companies.

component. One of the primary functional areas of an SRS. There are four surveillance components: Online Water Quality Monitoring (including source water and distribution system monitoring), Enhanced Security Monitoring, Customer Complaint Surveillance, and Public Health Surveillance. There are two response components: Consequence Management and Sampling and Analysis.

constraints. Requirements or limitations that may impact the viability of an alternative. The primary constraints for an SRS project are typically schedule, budget, and policy issues (for example, zoning restrictions, IT restrictions, and union prohibitions).

control center. A utility facility that houses operators who monitor and control treatment and distribution system operation, as well as other personnel with monitoring or control responsibilities. Control centers often receive system alerts related to operations, water quality, security, and some of the SRS surveillance components.

coverage. This is the availability of the technology at an SRS component site. Coverage will vary by location and technology. In general, the coverage of emerging technologies will expand as providers upgrade their infrastructure, while the coverage of older technologies will shrink as providers decommission aging infrastructure and phase-out obsolete systems.

cybersecurity. Measures implemented to protect an information management system and network from unauthorized access, damage, or attack. Common examples include password protected computers, encryption, and use of anti-virus software.

dashboard. A visually-oriented user interface that integrates data from multiple SRS components to provide a holistic view of distribution system water quality. The integrated display of information in a dashboard allows for more efficient and effective management of water quality and the timely investigation of water quality anomalies.

data directionality. See one and two-way communications. Data should flow from the remote site to the utility control center, and for certain applications should also be able to flow in the opposite direction for remote control functions.

data generation rate. The instantaneous rate of data produced by a piece of SRS equipment, such as a chlorine analyzer or video camera. This metric is typically expressed as data bits per second.

data management. The process of capturing, processing, and storing data.

data transmission rate. The maximum instantaneous rate of data that a technology can transmit. This metric is typically expressed as giga, mega or kilobits per second.

data packet size. SRS equipment will usually combine its generated data bits into a package of multiple bits, called a data packet, which is transmitted to its destination.

datastream. A time series of values for a unique parameter or set of parameters. Examples of SRS datastreams include, chlorine residual values, water quality complaint counts, and number of emergency department cases.

distance. The maximum distance that a technology can reliably transmit a packet of data without the aid of a signal amplifier. This metric depends on the transmitter signal strength, receiver sensitivity, physical media used for transmission and signal frequency (copper, fiber optic cable, or air).

Distributed Network Protocol 3 (DNP3). A communications protocol for real-time data transfer among programmable logic controllers and industrial electronic devices. DNP3 supports encryption and is capable of buffering data during communications outages and timestamping data.

Enhanced Security Monitoring (ESM). One of the surveillance components of an SRS. ESM includes the equipment and procedures used to detect and respond to security breaches at distribution system facilities that are vulnerable to contamination.

Equivalent Isotropically Radiated Power (EIRP). A commonly used measure of emitted radio frequency power referenced to milliwatts, is limited to 36 Decibels referenced to one milliwat (dBm).

extent of use. The measure of acceptance of a communications technology by utilities.

hardware. Physical IT assets such as servers or user workstations.

implementation costs. Costs to procure and install equipment, IT components, and other assets necessary to build an operational system.

information management system. The combination of hardware, software, tools, and processes that collectively support an SRS and provides users with information needed to monitor real-time system conditions. The system allows users to efficiently identify, investigate, and respond to water quality incidents.

information technology (IT). Hardware, software, and data networks that store, manage, and process information.

Internet of Things. The network of physical objects or "things" embedded with electronics, software, sensors, and connectivity to enable objects to exchange data with the manufacturer, operator and/or other connected devices based on the infrastructure of International Telecommunication Union's Global Standards Initiative. The Internet of Things allows objects to be sensed and controlled remotely across existing network infrastructure, creating opportunities for more direct integration between the physical world and computer-based systems, and resulting in improved efficiency, accuracy and economic benefit. Each thing is uniquely identifiable through its embedded computing system but is able to interoperate within the existing Internet infrastructure.

latency. The amount of time between when a data packet enters one end of a link and emerges from the other end. Latency is sometimes presented as "round-trip latency", which is the time between when a packet enters one end of a link and an acknowledgment is received back at the same end of the link. Latency is a function of data transmission rate, hardware and software processing time and the time required to establish a link.

lifecycle cost. The total cost of a system, component, or asset over its useful life. Lifecycle cost includes the cost of implementation, operation and maintenance, and renewal.

Long-Term Evolution (LTE). A standard for wireless communication of high-speed data for mobile phones and data terminals. LTE includes adaptive modulation technology to provide downloads and uploads of up to 300 and 75 megabits per second, respectively.

machine-to-machine communications. Technologies that allow both wireless and wired systems to communicate with other devices of the same type. This is a general concept that does not pinpoint specific wireless or wired networking, information and communications technology. Machine-to-machine communications is considered an integral part of the Internet of Things and brings several benefits to industry and business in general as it has a wide range of applications such as industrial automation, logistics, Smart Grid, Smart Cities, health, defense, etc. mostly for monitoring but also for control purposes.

maintenance. The cost and level of effort required to sustain the operation of a communication system.

modbus. A communications protocol for real-time data transfer among programmable logic controllers and industrial electronic devices.

one-way communications. A communications path that only allows the flow of data in one direction. Also referred to as uni-directional or simplex communications.

Online Water Quality Monitoring (OWQM). One of the surveillance components of an SRS. OWQM utilizes data collected from monitoring stations that are deployed at strategic locations in a source water or a distribution system. Monitored parameters can include common water quality parameters (e.g., pH, specific conductance, and turbidity) and advanced parameters (e.g., total organic carbon and spectral absorbance). Data from monitoring locations is transferred to a central location and analyzed for water quality anomalies.

operations and maintenance (O&M) costs. Expenses incurred to sustain operation of a system at an acceptable level of performance. O&M costs are typically reported on an annual basis, and include labor and other expenditures, such as supplies and purchased services.

polling interval. The frequency at which data is collected, reported, or transmitted.

provider fees. Fees to third-party communications provider. Fees can be assessed on a one-time basis such as installation charges for hardware implementation and configuration or recurring (e.g., monthly).

real-time. A mode of operation in which data describing the current state of a system is available in sufficient time for analysis and subsequent use to support assessment, control, and decision functions related to the monitored system.

reliability. Reliability means data reaches its destination completely, uncorrupted, and in the order it was sent.

round-trip latency. See Latency.

security. Although security has a broad definition that can vary among industries and technologies, for this document, security refers to hardware and software that ensures that the data transmitted between the remote and central sites cannot be viewed or altered by unauthorized personnel.

Service-level agreement. An agreement that defines uptime and reliability requirements, the timeline for the provider to address unscheduled outages, and penalties for the provider to pay when outages exceed the limits established in the agreement.

Smart Cities. Cities that use digital technologies or information and communication technologies to enhance quality and performance of urban services, to reduce costs and resource consumption, and to engage more effectively and actively with its citizens. Sectors that have been developing smart city technology include government services, transport and traffic management, energy, health care, water, and waste. Smart city applications are developed with the goal of improving the management of urban flows and allowing for real-time responses to challenges.

software. A program that runs on a computer and performs certain functions.

Supervisory Control and Data Acquisition (SCADA). A system that collects data from various sensors at a drinking water treatment plant and locations in a distribution system, and sends this data to a central information management system.

Third-party provided service. A company or organization that a utility would pay on a recurring basis for use of the provider's communications system. Typically third-party providers include the local telephone company, cable television providers and cellular carriers.

two-way communications. A communications path that allows the flow of data in both directions. Also referred to as bi-directional or duplex communications. Communications systems that allow simultaneous data flow in both directions are called full-duplex – devices on each end of the link are capable of talking and listening at the same time. Communications systems that allow data flow in both directions but only in one direction for a given instance are called half-duplex – devices on each end of the link are capable of talking and listening, but cannot do both at the same time.

uptime. The quantity of time when a communications system is performing reliably. See also **reliability**.

useful life. The period of time that an asset is able to be economically maintained.

user interface. A visually oriented interface that allows a user to interact with an information management system. A user interface typically facilitates data access and analysis.

water quality incident. An incident that results in an undesirable change in water quality (e.g., low residual disinfectant, rusty water, taste & odor, etc.). Contamination incidents are a subset of water quality incidents.

Water Quality Surveillance and Response System (SRS). A system that employs one or more surveillance components to monitor and manage source water and distribution system water quality in real time. An SRS utilizes a variety of data analysis techniques to detect water quality anomalies and generate alerts. Procedures guide the investigation of alerts and the response to validated water quality incidents that might impact operations, public health, or utility infrastructure.

Wi-Fi. Wireless local area network products that are based on the Institute of Electrical and Electronics Engineers' 802.11 standards. It is currently available in five different versions, 802.11b, 802.11g, 802.11n, 802.11ac, and 802.11ad which have data transmission rates of up to 11, 54, 450, 1,350, and 7,000 megabits per second, respectively.

WiMAX. Wireless local area network products that are based on the Institute of Electrical and Electronics Engineers' 802.16e standards. It stands for Worldwide Interoperability for Microwave Access and is designed to provide 30 to 40 megabits per second data rates.

wired communications. A method of transmitting data that uses a solid material such as copper or fiber optic cabling as the physical media.

wireless communications. A method of transmitting data that uses air as the physical media.

Appendix: Calculating Data Generation

Although empirical data is preferred to estimate data generation rates and the data generated per month from SRS equipment, such data may not be available or feasible to collect. Data generation rates may need to be calculated to ensure that a communications system has adequate capacity to transmit data from SRS equipment. Furthermore, estimated data generation rates can be used by a utility to budget for anticipated recurring costs when using a third-party communications provider. The methods for estimating the data generation rates and data generated per month for ESM non-video, ESM video, and OWQM water quality parameter data are described below.

A.1 Enhanced Security Monitoring Non-Video

Without video transmission, the overall data generation rate of ESM is relatively small because alerts occur intermittently and consist of a data packet that is on the order of a kilobit. The data generated per month of ESM will depend on the number of alerts generated.

A.1.1 Approximate the Maximum Data Generation Rate for Each Remote Site

Many third-party communications providers price their data services based on maximum data transmission rate capacity, which is the most bits of data per second that their service will support. For a communications application to work properly, the data generation rate should be less than the data service's data transmission rate. The data generation rate demand of a typical application varies with user-initiated and programmatic data transmissions. The generation rate can range from near zero when user activity is minimal and programs are dormant to the maximum generation rate when peak levels of data are produced. Non-video monitoring applications have relatively low data generation rates, and this metric should be approximated using the approach described below such that the appropriate data service can be evaluated.

For utility-owned options, the maximum data generation rate can be used when evaluating hardware for a new communications system or reviewing the existing communications infrastructure for adequate capacity. For existing communications systems, a small-scale pilot may be useful for empirically determining the system's suitability for transmitting ESM data.

Most non-video monitoring systems transmit data only after an intrusion has been detected with only minimal overhead data traffic (e.g., a few data bits to confirm that the communications link is still functioning) during non-alert conditions.

The data generation rate for non-video data depends on the transmission protocol, encryption, and polling method. Each of these variables is explained below. Data generation rates are usually presented as bits per second.

- **Transmission Protocol:** In the water sector, two protocols are typically used for data transmission from remote sites: ***Modbus*** and ***Distributed Network Protocol 3 (DNP3).*** Modbus is used in many legacy systems, but newer installations have been using DNP3 because of key features not available in Modbus such as timestamping and store-and-forward ability. The added functionality of DNP3 also means that this protocol requires more data overhead than Modbus, although at the data generation rates of non-video ESM applications, the additional overhead is usually not a limitation.

- **Encryption:** For an added level of cybersecurity, data can be encrypted, although encryption adds additional overhead data to the transmitted data packets. The two most common methods of encryption are AES-128 and AES-256, which add 128 and 256 bits of overhead per data packet, respectively.

- **Polling Method:** Non-video ESM data usually consists of discrete alert signals, which can be transmitted when a change-of-state in one of the signals occurs or periodically based on a ***polling interval*** (host-polled). The data generation rate and data generated per month calculations vary between the two methods and are described below.

<u>Data Generation Rate Approximation for a Host-Polled System</u>

The data generation rate calculation described below assumes that a host-polled system is being used, the entire data packet is received before the next poll, and polls are successful 50% of the time to account for packets that are lost in transmission. Determining a polling interval that will satisfy the ESM component's operational needs is also required for this calculation. Third-party communications systems can typically accommodate short polling intervals such as once every 2 seconds, although point-to-multi-point utility-owned wireless solutions may have polling intervals on the order of minutes.

The following data generation rate formulas are presented for the DNP3 protocol for a conservative estimate of the data generation rate because it has more overhead than most legacy protocols such as Modbus.

$$\text{Data Rate} = \frac{\text{Data Packet Size}}{\text{Polling Interval} \div 2}$$

$$\text{Data Packet Size} = [\text{Overhead} + \text{Encrypt} + (\text{Discrete Points} \times B)]$$

Where:

Data Rate =	Data Generation Rate (bits per second)
Overhead =	Bits of protocol-specific overhead (656 for DNP3)
Encrypt =	Encryption bits (128 for AES-128 encryption, 256 for AES-256, or 0 if unencrypted)
Discrete points =	Number of discrete alert signals from the ESM station, such as door, motion sensor, and other intrusion detection alerts
B =	Bits per discrete point (56 for DNP3)
Polling Interval =	Amount of time between data polls in seconds

Equation A-1. Data Generation Rate – Host-Polled

Example – Data Generation Rate (Host-Polled)

The input parameters for this example are as follows:

- **Discrete Points:** 4 (Entry Door Alert, Motion Sensor Alert, Motion Sensor Fault, Pump Room Door Alert). These binary-type alert or status signals are typically generated by intrusion detection equipment at an ESM site. Some sensors transmit a fault or "service needed" signal if an internal fault is detected. Consult the sensor manufacturer for the available discrete signals.

- **Polling Interval:** Data requested every 5 seconds.

- **Encryption:** Data encrypted with AES-256. This is considered a secure means of encrypting data, which is recommended for critical infrastructure applications.

Given the above parameters, the data generation rate calculation is as follows:

$$\textbf{Data Packet Size = [(656+256)+(4×56) bits] = 1,136 bits}$$

$$\textbf{Data Rate} = \frac{\textbf{1,136 bits}}{\textbf{5 seconds} \div 2} = \textbf{454 bits per second}$$

Data Generation Rate Approximation for a Report-By-Exception System

The data generation rate calculation described below assumes that data is only generated when an alert occurs and that only data for the active alert is transmitted.

The following data generation rate formulas are presented for the DNP3 protocol for a conservative estimate of the data generation rate because it has more overhead than most legacy protocols such as Modbus.

$$\textbf{Data Rate} = \frac{\textbf{Data Packet Size}}{\textbf{Transmission Time}}$$

$$\textbf{Data Packet Size = [Overhead+Encrypt+B+TA]}$$

Where:
Data Rate =	Data Generation Rate (bits per second)
Overhead =	Bits of protocol-specific overhead (656 for DNP3)
Encrypt =	Encryption bits (128 for AES-128 encryption, 256 for AES-256, or 0 if unencrypted)
B =	Bits per discrete point (56 for DNP3)
TA =	Bits per point tag (192 for DNP3); this indicates which point is alerting
Transmission Time =	Desired amount of time for the alert to be transmitted from the remote site to the host computer in seconds

Equation A-2. Data Generation Rate – Report by Exception

Example – Data Generation Rate (Report by Exception)
The input parameters for this example are as follows:

- **Transmission Time:** The user wants alerts to take no more than 7 seconds to arrive at the control center.

- **Encryption:** Data encrypted with AES-256. This is considered a secure means of encrypting data, which is recommended for critical infrastructure applications.

Given the above parameters, the data generation rate calculation is as follows:

$$\textbf{Data Packet Size = [656 bits+256 bits+56 bits+192 bits] = 1,160 bits}$$

$$\textbf{Data Rate} = \frac{\textbf{1,160 bits}}{\textbf{7 seconds}} = \textbf{166 bits per second}$$

A.1.2 Approximate the Data Generated Per Month for Each Remote Site

Some third-party communications providers, such as cellular carriers, offer data plans that are based on a monthly data allotment where overage charges are incurred if the monthly data limit is exceeded. The plan may be based on a per site basis or other negotiated terms. If such a plan is being considered, the following data generated per month formulas can be used to approximate the amount of data from each remote site over a month such that the appropriate data plan can be evaluated. Key inputs to this calculation include the polling interval (if using a host-polled system), as previously determined for Equation A-1, and an estimate of the number of alerts per month. Data generated per month is typically presented as bytes. (Note: There are 8 bits in a byte, and the constant "8" is a conversion factor in the formulas below.)

Data Generated Per Month Approximation for a Host-Polled System

The following formula can be used to calculate the data generated per month for host-polled systems. Key inputs to this calculation include the polling interval and Data Packet Size, as previously shown in Equation A-1.

Data Generated = [(Poll Packet Size)+(Data Packet Size)]×Polls Per Month

Where:
Data Generated =	Data generated per month (bytes)
Poll Packet Size =	[Overhead + Encrypt] (from the Data Generation Rate Formula, divided by 8) + 2 bytes
Data Packet Size =	As calculated in the Data Generation Rate Formula (bits), divided by 8
Polls Per Month =	Number of polls per month

Equation A-3. Data Generated Per Month – Host-Polled

Example – Data Generated Per Month (Host-Polled)
The input parameters for this example are the same as those used in the previous example. Specifically, the overhead + encryption value was calculated as 912 bits (656 + 256) and the data packet size, 1,136 bits. Thus, the data generated per month calculation is as follows:

Poll Packet Size = (912÷8)+2 = 116 bytes

Data Packet Size = 1,136÷8 = 142 bytes

Polls Per Month = (1 poll /5 seconds)×(2,628,000 seconds/month) = 525,600 polls/month

Data Generated = [116+142]×525,600 = 135,604,800 bytes per month

Data Generated Per Month Approximation for a Report-By-Exception System

The following formula can be used to calculate the data generated per month for report-by-exception systems. Key inputs to this calculation include the estimated number of alerts per month and Data Packet Size, as previously shown in Equation A-2.

Data Generated = [Data Packet Size]×Alerts Per Month

Where:
Data Generated =   Data generated over a month (bytes)
Data Packet Size =   As calculated in the Data Generation Rate Formula (bits per second),
                       divided by 8
Alerts Per Month =   Number of alerts per month

Equation A-4. Data Generated Per Month – Report-By-Exception

Example – Data Generated Per Month (Report-By-Exception)
The input parameters for this example are the same as those used in the previous example. Specifically, the overhead + encryption value was calculated as 912 bits (656 + 256) and the data packet size, 1,160 bits. Additionally, 10 alerts are estimated per month. Thus, the data generated per month calculation is as follows:

Data Packet Size = 1,160 ÷8 = 145 bytes

Alerts Per Month = 10

Data Generated = 145 bytes per alert ×10 alerts per month = 1,450 bytes per month

A.2   Enhanced Security Monitoring Video

ESM video applications can utilize a significant amount of bandwidth depending on the characteristics of the imagery, typically on the order of megabits per second for continuous video. Estimating the data generation rate is critical to determine if the communications technology has adequate transmission capacity for this data-intensive application.

A.2.1 Approximate the Maximum Data Generation Rate for Each Remote Site

Many third-party communications providers price their data services based on maximum data transmission rate capacity, which is the most bits of data per second that their service will support. For a communications application to work properly, the data generation rate should be less than the data service's data transmission rate. The data generation rate demand of a typical application varies with user-initiated and programmatic data transmissions. The generation rate can range from near zero when user activity is minimal and programs are dormant to the maximum generation rate when peak levels of data are produced. Video monitoring applications typically have high data generation rates, and this metric should be approximated using the approach described below such that the appropriate data service can be evaluated.

For utility-owned options, the maximum data generation rate can be used when evaluating hardware for a new communications system or reviewing the existing communications infrastructure for adequate capacity. For existing communications systems, a small-scale pilot may be useful for empirically determining the system's suitability for transmitting video data traffic.

Most video monitoring systems can operate in a continuous or incident-based mode, and each mode can have a different data generation rate. Thus the maximum data generation rate should be the higher of the two. When operating in incident-based mode, the video system does not transmit data under non-alert conditions and only transmits a video clip of predefined duration when an intrusion has been detected by a sensor connected to the video system. The time to transmit the video clip can be longer than the duration of the clip itself to reduce the data generation rate. Most video systems with incident-based capability also allow the user to switch to continuous mode, which is useful when utility personnel want a ***real-time*** view of a facility (e.g., immediately after an intrusion has occurred), but this can result in a relatively large data generation rate.

The data generation rate for incident-based or continuous video depends on their respective image resolution, frame rate, intraframe compression, and interframe compression. The settings for these variables are usually independently adjustable, resulting in different data generation rates for the two modes of operation, and the higher of the two should be used as the maximum. Each of the variables are explained below. Data generation rates are usually presented as bits per second.

- **Image Resolution:** Image resolution is the number of pixels in an image, and selecting a resolution typically balances the intended use of the imagery with the camera cost. For example, if a utility requires that personnel be able to recognize faces in the video image, a high resolution will be required, and the cameras will be more costly. However, if a utility wants to be able to discern whether the person in the image is wearing a utility uniform, then a lesser resolution (and less expensive) camera can be used. For facial recognition, the general resolution guideline is approximately 80 pixels per foot. Thus if a camera is monitoring a 10-foot wide area, the camera will need a resolution of at least 800 pixels wide for facial recognition. Therefore, a camera having a resolution of Super Video Graphics Array (SVGA), which is 800 x 600 pixels, might be considered. For general recognition, approximately 30 pixels per foot is recommended. For the 10-foot wide area, an image 300 pixels wide would be necessary for general recognition purposes, so a camera having a resolution of Common Intermediate Format, which is 352 x 288 pixels, would be adequate. However, the cost difference between SVGA and Common Intermediate Format cameras is low, and furthermore, it may be difficult to find a current IP camera with a resolution of only Common Intermediate Format. High-end high definition models have a resolution of 1920 x 1080 pixels. Another consideration is the advantage of providing consistent camera models across the entire facility, lowering costs for spares. Generally, if a higher resolution is desired, a higher data generation rate is produced.

- **Frame Rate:** Frame rate is the frequency (rate) at which an imaging device produces unique consecutive images. Cameras with frames rates up to 30 frames per second are available, but 12 to 15 frames per second can provide adequate image capture for many applications. If video analytic software will be used, higher frame rates may be required. Generally, if more frames per second are desired, the data generation rate is higher.

- **Intraframe Compression:** Intraframe compression is the amount that an individual frame can be compressed before noticeable detail is lost. Compression depends on the variability of color and composition in the image. Images that have large spaces with similar or identical colors will be more compressible and thus generate less data when compared with images that include numerous objects and colors. Intraframe compression is used to create standalone frames called "I-frames."

- **Interframe Compression:** Video compression algorithms compress a series of frames by only transmitting what has changed from one frame to the next, instead of sending complete images for each frame. Each frame that only consists of data that has changed since the previous frame is called a "P-frame." This method drastically reduces the amount of data being transmitted while maintaining high-quality imagery. Video compression algorithms control the number of P-frames that are used between I-frames. Scenes with minimal motion, such as the door to an unstaffed facility, can produce high levels of intraframe compression due to relatively small P-frames. Conversely, scenes with many moving objects, such as a train station, will have larger P-frames and be less compressible on an intraframe basis. Larger P-frames will also occur for scenes with pixel variability produced by the camera, which often results when the camera's electronic eye attempts to resolve scenes with insufficient lighting. Any variability in an image will be considered as a change by the video algorithm and result in poor intraframe compression even if there is minimal actual motion. A camera designed for low-light conditions for such an application should be considered to minimize image variability and improve compression levels. Generally, if images differ greatly due to movement or are captured with cameras not optimized for lighting conditions, a higher data generation rate may result.

<ins>Data Generation Rate Approximation for Continuous Video</ins>

It is strongly recommended that the user obtain camera-specific information from the manufacturer for estimating the data generation rate required of the video application. Most camera manufacturers provide data calculators on their websites to demonstrate how compression, resolution, and frame rate affect the image quality and data generation rate for different scene types. Users should consult these online resources to approximate data generation rates for their applications, keeping in mind that compression levels are specific to the camera manufacturer, so results from one manufacturer's data calculator may not correspond to another's for compression. However, frame rates and resolutions are standardized across the video industry. Furthermore, empirically determining a camera's data generation rate for different scene-types using a small-scale pilot at an actual utility facility can be a useful exercise for confirming camera-specific data generation rates.

As a point of reference, the uncompressed data generation rate for an SVGA video feed at 12 frames per second is approximately 69 megabits per second, but, even with a minimal level of compression, a multicolored scene with many moving objects at this resolution and frame rate can be reduced to approximately 2.5 megabits per second. At this level of compression, images remain clear with minimal pixelation.

Scene type significantly affects compression effectiveness, and a scene with minimal motion and large areas of similar color can greatly improve compression. Such a scene using the same level of compression, frame rate, and resolution in the previous example can result in a data generation rate of approximately 0.4 megabits per second. Generally, higher levels of compression can be applied to reduce the data generation rate, but image quality may be impacted.

<ins>Data Generation Rate Approximation for Video Clips</ins>

First, the continuous data generation rate associated with the video clip should be approximated by inputting the video clips' frame rate, resolution, and compression level into the manufacturer's data calculator. Next, the following equation can be used to determine the data generation rate for transmitting a video clip.

$$\text{Video Clip Data Rate} = \frac{CR \times CD}{T}$$

Where:
CR = Video Clip Continuous Data Generation Rate (bits per second)
CD = Video Clip Duration (seconds)
T = Desired time for Utility personnel to receive the clip (seconds)

Equation A-5. Video Clip Data Generation Rate

Example – Data Generation Rate Approximation for Video Clips

A utility wants facial recognition from a camera that views the entry door at an unstaffed pump station. The view across the doorway is approximately 10 feet and has a moderate amount of color variation and objects near the doorway. The resolution required for facial recognition is 80 pixels per foot, so a minimum resolution of 800 x 600 (SVGA) is required. Using a manufacturer's data calculator, the utility inputs the SVGA resolution and a 12 images per second frame rate for smooth imagery. Based on sample videos provided by the manufacturer that demonstrate varying levels of compression and resolution, the utility estimates that a compression of 10 provides adequate image quality and selects a scene type that best represents the pump station. Given these input parameters, the data calculator yields a data generation rate of approximately 0.889 megabits per second if continuous video is used for this application. If this is too high, the user may consider reducing the frame rate as it is linearly related to the data generation rate (e.g., a frame rate of 6 frames per second will produce a data generation rate of 0.444 megabits per second). Using an incident-based video system is another option for reducing the data generation rate.

If the utility was considering an incident-based mode for the above application, the data generation rate would be calculated as follows. This example assumes that the frame rate and resolution are the same as those for the continuous mode, the video clip is 10 seconds, and the utility wants to receive the clip at its control center 30 seconds after the end of the clip – utility personnel see a 10-second video of the intrusion 40 seconds after the entry occurs.

$$\text{Video Clip Data Rate} = \frac{0.889 \text{ megabits per second} \times 10 \text{ seconds}}{30 \text{ seconds}}$$
$$= 0.296 \text{ megabits per second}$$

Where:
  CR = 0.889 megabits per second
  CD = 10 seconds
  T = 30 seconds

The higher of the continuous and incident-based video clip data generation rates should be used as the maximum value. For this example, the higher rate is 0.889 megabits per second, which was the continuous data generation rate. However, if the video clip resolution was significantly more than that of the continuous view, the incident-based video clip generation rate could be the higher value. For this example, POTS would not be a viable option, and 3G cellular technology and frequency-hopping spread spectrum transceiver (utility-owned) would be marginal. Thus, non-POTS wired technologies and wireless technologies that use adaptive modulation (e.g., 4G cellular) should be considered for this application.

A.2.2 Approximate the Data Generated Per Month for Each Remote Site

Some third-party communications providers, such as cellular companies, offer data plans that are based on a monthly data allotment where overage charges are incurred if the monthly data limit is exceeded. The plan may be based on a per site basis or other negotiated terms. If such a plan is being considered, the following Data Generated per Month equations can be used to approximate the amount of data from each remote site over a month such that an appropriate data plan can be evaluated. The calculation for a continuous-only video application (Equation A-6) is relatively straight forward, but estimating the data generated per month for an application with incident-based and continuous modes (Equation A-7) is more involved. Key inputs to this calculation include an estimate of the number of alerts per month, the duration of a video clip for an alert, the data generation rate of a video clip, the duration of continuous video viewing after an alert has occurred, the frame rate ratio between the video clip and live video, and the data generation rate of continuous video. The data generation rate for video clips and continuous video are differentiated to account for video systems that allow the user to independently adjust the resolutions and frame rates of video clips and live video. Both equations would be required if the utility has an application with multiple cameras where some operate in continuous mode while others are incident-based. Data generated per month is typically presented as bytes. (Note: There are 8 bits in a byte, and the constant "8" is a conversion factor in the formulas below.)

Data Generated Per Month for Continuous Video

The formula for calculating data generated per month for continuous video applications is as follows:

Data Generated Per Camera = CR ×SPM

SPM = 60×60×24×365÷12 = 2,628,000 seconds per month

Where:
CR = Continuous Video Data Generation Rate (from the manufacturer's data calculator in bits per second, divided by 8)
SPM = Seconds per month

Equation A-6. Data Generated per Month per Camera

Data Generated Per Month for Incident-Based and Continuous Video

The formula for calculating data generated per month for incident-based and continuous video applications is as follows:

Data Generated = [(CR ×CD)+ (LR ×LF ×LD)]×A

Where:
CR = Video Clip Data Generation Rate (megabits per second, divided by 8)
CD = Video Clip Duration (seconds)
LR = Live Video Data Generation Rate (megabits per second, divided by 8)
LF = Live Video Frame Rate Divided by the Video Clip Frame Rate
LD = Live Video Duration (seconds)
A = Estimated number of alerts per month

Equation A-7. Data Generated Per Month

Two key data considerations are downloading system updates and camera control. Users should consult the video system manufacturer for the approximate amount of data per update and an estimate of the number of downloads and updates that could occur in a month. If the video monitoring system has the ability to control pan-tilt-zoom cameras, the outgoing data associated with camera control should also be considered as some communications technologies, such as DSL, have different outgoing and incoming data capacities. Generally, the amount of data required for camera control should be relatively small compared to that of video data.

Example – Data Generation Rate Approximation for Incident-Based and Continuous Video
A camera monitoring the doorway from the previous example requires that the video clip provide facial recognition so that utility personnel can determine the person's identity as he or she enters the door. The continuous video feed requires only general recognition so that utility personnel can follow the intruder and determine intent, thus a higher level of compression can be applied. Based on the manufacturer's data calculator, this data generation rate is 0.514 megabits per second (compared to the 0.889 required for facial recognition). A frame rate of 12 per second is required of the video clip and 6 per second of the

continuous feed. Furthermore, the video clip duration is 10 seconds, and it is assumed that utility personnel will view the intrusion in continuous mode for 60 seconds. Lastly, two alerts per month are expected. The calculation is as follows:

$$\text{Data Generated} = [\left(\frac{0.889}{8} \times 10\right) + \left(0.514 \times \frac{6}{12} \times \frac{60}{8}\right)] \times 2 = 6.1 \text{ megabytes per month}$$

Where:
CR = 0.889 / 8 (0.889 is from the manufacturer's data calculator in the first example)
CD = 10 (seconds)
LR = 0.514 / 8 (0.514 is from the manufacturer's data calculator, divided by 8)
LF = 6 / 12
LD = 60 seconds
A = 2 alerts per month

A.3 Water Quality Parameter Data

An OWQM station that transmits only parameter data typically has a data generation rate on the order of kilobits per second depending on the number of parameter values. The data generated per month of an OWQM station will depend on the frequency of data transmissions. For OWQM stations with spectral data, the data generation rate and data generated per month can be significantly more than that of parameter-only transmissions, and the utility should consult the instrument manufacturer for details.

A.3.1 Approximate the Maximum Data Generation Rate for Each Remote Site

The data generation rate calculation described below assumes that water quality data is periodically transmitted to a utility control center (i.e., host-polled, as previously defined for ESM), each data transmission is completed before the next transmission begins, and transmissions are successful 50% of the time to account for lost data. Determining a data transmission frequency (also called a polling interval) that will satisfy the OWQM component's operational needs, typically on the order of every 2 to 5 minutes, is also required for this calculation. Consult *Online Water Quality Monitoring Design Guidance For Distribution System Monitoring* for details on determining the polling interval for an OWQM application. Data generation rates are usually presented as bits per second.

The following Data Generation Rate Formula is presented for the DNP3 protocol, which originated in the power industry but is gaining acceptance in the water sector due to key features such as time-stamping and store-and-forward capabilities. From a data generation rate perspective, the DNP3 protocol has more overhead than most legacy protocols such as Modbus, so assuming DNP3 usage provides a conservative estimate of the data generation rate.

$$\text{Data Rate} = \frac{\text{Data Packet Size}}{\text{Poll Time} \div 2}$$

Data Packet Size = [Overhead+Encrypt+(Analog Points)×A+(Discrete Points)×B]

Where:
Data Rate =	Data Generation Rate (bits per second)
Overhead =	Bits of protocol-specific overhead (656 for DNP3)
Encrypt =	Encryption bits (128 for AES-128 encryption, 256 for AES-256, or 0 if unencrypted)
Analog points =	Number of analog datastreams from the OWQM station, such as pH, TOC, etc.
A =	Bits per analog point (72 for DNP3)
Discrete points =	Number of discrete status/alert points from the OWQM station, such as instrument faults, alerts, etc.
B =	Bits per discrete point (56 for DNP3)
Poll Time =	Amount of time between data polls in seconds

Equation A-8. OWQM Data Generation Rate

Example

Input Parameters:

- **Analog Points:** 5 (assume pH, TOC, temperature, conductivity, and free chlorine). These analytical readings are from typical water quality sensors at an OWQM site.

- **Discrete Points:** 3 (instrument fault, alert, error). These binary-type alert or status signals are typically generated by water quality sensors or anomaly detection systems at an OWQM site. Most sensors transmit a fault or "service needed" signal if it requires more reagents, is due for routine maintenance, or has detected an internal problem. If an OWQM site includes an anomaly detection system, it will generate an alert signal when anomalous water quality has been detected or an error signal if the device has experienced an internal failure.

  Consult the sensor manufacturer for the available analog and discrete signals.

- **Polling Interval:** Data requested every 5 seconds.

- **Encryption:** Data encrypted with AES-256. This is considered a secure means of encrypting data, which is recommended for critical infrastructure applications.

Given the above input parameters, the data packet size and data rate calculations are as follows:

Data Packet Size = [(656+256)+5×72 bits+3×56 bits]= 1,440 bits

$$\text{Data Rate} = \frac{\textbf{1,440 bits}}{\textbf{5 seconds} \div 2} = \textbf{576 bits per second}$$

A.3.2 Approximate the Data Generated Per Month for Each Remote Site

The following formula can be used to calculate the data generated per month for host-polled systems. Key inputs to this calculation include the polling interval and Data Packet Size, as previously calculated in Equation A-8. Some host-polled systems also allow for alerts and status signals to be transmitted immediately upon change-of-state without waiting for the next poll to occur. Typically over a month, the amount of alert data generated on a change-of-state basis is relatively small compared to the data generated from polled transmissions. In most cases, the polling interval required for OWQM applications are frequent enough such that non-polled alert transmission is not needed. (Note: There are 8 bits in a byte, and the constant "8" is a conversion factor in the formulas below.)

Data Generated = [(Poll Packet Size)+(Data Packet Size)]×Polls Per Month

Where:

Data Generated =	Data generated over a month (bytes)
Poll Packet Size =	[Overhead + Encrypt] (from the Data Generation Rate Formula, divided by 8) + 2 bytes
Data Packet Size =	As calculated in the Data Generation Rate Formula, divided by 8
Polls Per Month =	Number of polls per month

Equation A-9. OWQM Data Generated Per Month

For applications where the user can download data or update firmware over a communications link, consult the instrument manufacturer for the approximate amount of data per download or update and an estimate of the number of downloads and updates that could occur in a month.

Example

Input Parameters:

- Same as for the previous example. Specifically, the overhead + encryption value was calculated as 912 bits (656 + 256) and the data packet size, 1,440 bits.

Poll Packet Size = (912 ÷8)+2 = 116 bytes

Data Packet Size = 1,440 ÷8 = 180 bytes

Polls Per Month = (1 poll /5 seconds)×(2,628,000 seconds/month) = 525,600 polls/month

Data Generated = [116+180]×525,600 = 155,577,600 bytes per month

A.4   Summary of Input Parameters

A summary of the component-level inputs needed for calculating data rates and usage for each of the above applications is shown in **Table A-1**.

Table A-1. Summary of Component-Level Inputs for Calculating Data Rates and Usage

Component Application	Input Needed	Comments
ESM Non-Video	Quantity of alert points at a facility	Assumes that the status values of all alerts will be sent in a single data packet.
	Transmission protocol	Modbus and DNP3 are typical choices.
	Level of encryption	128 or 256-bit are typical choices.
	Polling method	Host-polled or report-by-exception.
	Polling interval	Applies to a host-polled system.
	Desired transmission time for alerts to arrive	Applies to a report-by-exception system.
	Number of alerts received per month	Needed to calculate monthly data usage for a report-by-exception system.
ESM Video	Image resolution	Use the recognition guidelines described in Section A.2.1.
	Frame rate	Use the sample video from the camera manufacturer's data calculator to determine an adequate frame rate.
	Compression level	Use the sample video from the camera manufacturer's data calculator to determine an adequate compression level.
	Video clip duration	Applies to an incident-driven video monitoring system.
	Desired transmission time for alerts to arrive	Applies to an incident-driven video monitoring system.
	Number of alerts received per month	Needed to calculate monthly data usage for an incident-driven video monitoring system.
OWQM (non-spectral)	Quantity of analog and alarm points at a facility	Assumes that the values of all water quality parameters and alarm statuses will be sent in a single data packet.
	Transmission protocol	Modbus and DNP3 are typical choices.
	Level of encryption	128 or 256-bit are typical choices.
	Polling interval	Assumes a host-polled system.